WALKING TO OLYMPUS

AN EVA CHRONOLOGY, 1997–2011

VOLUME 2

Julie B. Ta
Robert C. Treviño

MONOGRAPHS IN AEROSPACE HISTORY SERIES #50
APRIL 2016

National Aeronautics and Space Administration

NASA History Program Office
Public Outreach Division
Office of Communications
NASA Headquarters
Washington, DC 20546

NASA SP-2016-4550

Library of Congress Cataloging-in-Publication Data

Ta, Julie B., author.
Walking to Olympus: an EVA chronology, 1997–2011 / by Julie B. Ta and Robert C. Treviño. – Second edition.
pages cm. – (Monographs in aerospace history series; #50)
"April 2016."
Continuation of: Walking to Olympus / David S.F. Portree and Robert C. Treviño. 1997.
"NASA SP-2015-4550."
Includes bibliographical references and index.
1. Extravehicular activity (Manned space flight)–History–Chronology.
I. Treviño, Robert C., author. II. Title.
TL1096.P67 2015
629.45'84–dc23
2015030907

ON THE COVER

Astronaut Steve Robinson, anchored to a foot restraint on the International Space Station's Canadarm2, participates in the STS-114 mission's third spacewalk. Robinson holds a digital still camera, updated for use on spacewalks, in his left hand. (NASA S114e6651)

This publication is available as a free download at *http://www.nasa.gov/ebooks.*

ISBN 978-1-62683-031-8

CONTENTS

FOREWORD

As former NASA Chief Historian Roger Launius noted in the foreword to the first edition of this chronology, one of the most significant activities conducted in space happens when humans leave their spacecraft and operate in spacesuits. Extravehicular activities (EVAs) require great technical skills, sophisticated technologies, meticulous planning and training, and careful coordination of human capabilities. EVAs have not only accomplished significant work in space that could not be done any other way, but they have also yielded enormous knowledge, skills, and experience among the astronaut and cosmonaut corps about how to work beyond the confines of Earth's atmosphere.

As somebody who has designed spacesuits, pushed technology innovation, and devoted considerable time to thinking about how humans can best perform and use tools in space, this publication is intensely interesting to me. Human spaceflight is often considered the greatest challenge in space exploration and EVAs are at the crux of human spaceflight and exploration. While there are still many uncertainties about our eventual human voyages to Mars, there is still at least one certainty: we will need to be able to work successfully outside of the space vehicles and habitats to make that journey. In doing so, we will inevitably build on the rich history of EVAs that has enabled us to walk on the Moon, service the Hubble Space Telescope, build the International Space Station, and much more. By understanding our past successes and limitations, we can ensure that the best is indeed yet to come.

Dr. Dava J. Newman
Deputy Administrator
National Aeronautics and Space Administration

14 July 2015

INTRODUCTION

The first edition of *Walking to Olympus: An EVA Chronology* (Monograph in Aerospace History #7, October 1997) spanned a period of space exploration of 32 years, from the first spacewalks in 1965 to the end of the Shuttle-Mir program in 1997. It included EVAs performed by both Soviet/Russian cosmonauts and American astronauts. The Soviet/Russian space programs that involved spacewalks were the Voskhod, Soyuz, Salyut, Mir, and Shuttle-Mir. During this same time period, the USA space programs that included spacewalks were Gemini, Apollo, Skylab, Space Shuttle, and the Shuttle-Mir.

This second volume of *Walking to Olympus* continues from the end of the Shuttle-Mir program in 1997 to the end of the Space Shuttle Program in 2011. It includes not only spacewalks performed by American and European astronauts and the Russian/Soviet cosmonauts, but also those of the newest members of the EVA community, the taikonauts of the People's Republic of China (Chinese taikonauts performed their first spacewalk on 27 September 2008). Space programs with EVAs that are included in this second volume are: the Mir and the International Space Station (ISS) programs (Russia), the Space Shuttle and the ISS programs (USA), and the Shenzhou space program (China).

The publication of this second EVA chronology of spacewalks coincides with two major anniversaries of significance to the human spaceflight community: the 50th Anniversary of EVA and the 25th anniversary of the Hubble Space Telescope (HST). The HST was launched on STS-31 on 24 April 1990. Repaired, maintained, and upgraded by EVA crewmembers over five Hubble Servicing Missions spanning 16 years, HST has continued to give the world a whole new view and understanding of our universe thanks to EVA technology, and the hard work of the spacewalkers.

Several key events and themes are notable during the time period this monograph covers: the building of the ISS, the servicing of the HST, and the STS-107 Columbia accident. For those interested in understanding how spacewalkers assembled the various modules of ISS, please see Gary Kitmacher's *Reference Guide to the International Space Station* (NASA SP-557, 2006). Several other publications cover various aspects of ISS history, and we also recommend *http://spaceflight.nasa.gov/history/station/index.html* for more information. Regarding HST history, there is a selective bibliography online at *http://history.nasa.gov/hubble/bib.html.* The scholarly starting point on this subject is Robert Smith's *The Space Telescope: A Study of NASA, Science, Technology, and Politics* (Cambridge University Press 1989), which takes the story up to launch; a follow-up volume covering HST's operations is being researched and written now. Roger Launius and David DeVorkin's *Hubble's Legacy: Reflections by Those Who Dreamed It, Built It, and Observed the Universe with It* (Smithsonian Press 2014) is an excellent edited volume of relevant essays. The STS-107 Columbia accident on 1 February 2003, was certainly a major defining event for NASA in general and human spaceflight more specifically. For basic background on what happened and why, a good starting point is *http://history.nasa.gov/columbia*.

Spacewalks have become iconic and still capture the public's imagination when they see a human in a spacesuit in the extreme environment of space. The phrase "Walking to Olympus" is a symbolic expression of humans inevitably landing on Mars and exploring the planet for humanity, including Olympus Mons, the largest volcano in our solar system.

Roland W. Newkirk, Ivan D. Ertel, and Courtney G. Brooks's *Skylab: A Chronology* (NASA SP-4011, 1977) was the inspiration for this EVA chronology. Its Foreword, written by Skylab astronaut Charles Conrad, Jr., stated, "This chronology relates only the beginning; the best is yet to come from Skylab." *Walking to Olympus* also narrates only the beginning, as the best is yet to come from further developments of EVA technology, as well as the future discoveries and applications made possible by EVA.

In keeping with the writing style and tradition of the first volume, this second volume has maintained essentially the same format. Our readers can follow the information more easily by noting the definitions and criteria we established to maintain consistency in the EVA times, dates, names, and format of the text. These definitions include the following:

Starting with the Space Shuttle Program, EVA start and end times are based on the spacesuit's Portable Life Support System's Battery On and Battery Off times during an EVA.

Date of an EVA is the day that the EVA starts since several EVAs start in one day, go past midnight, and end the next day.

EVA crewmembers are listed in the order of lead spacewalker and then other EVA crewmembers whenever possible.

The authors gratefully acknowledge the support and encouragement by many in the NASA, contractor, and academia team that makes up the EVA community. The authors would like to acknowledge Bill Barry, NASA Chief Historian; Steve Garber of the NASA History Program Office; the NASA Johnson Space Center (JSC) History Office; the NASA JSC Mission Operations Directorate; the EVA Division that maintains the official times for the EVAs; the NASA JSC EVA Office that also maintains an EVA database that was used to cross-reference information; Richard Fullerton; Keith Johnson; Randall McDaniel; Jeannie Corte; the Crew and Thermal Systems Division; and the many contractors associated with EVA. We also thank the peer reviewers for their thoughtful feedback; as well as Betty Ta, King Ta, and Matthew Stuhl whose encouragement was invaluable.

Thanks are also due to a number of talented professionals who helped bring this project from manuscript to finished publication. Editor Yvette Smith in the NASA History Program Office performed her usual outstanding work preparing the manuscript for production. In the Communications Support Service Center: J. Andrew Cooke carefully copyedited the text, Michele Ostovar did an expert job laying out the design graphically and creating the e-book version, Kristen Harley thoroughly indexed our work, and printing specialist Tun Hla made sure the traditional hard copies were printed to exacting standards. Supervisors Barbara Bullock and Michele Ostovar helped by overseeing all of this CSSC production work.

Of course, the authors assume responsibility for any error.

THE CHRONOLOGY

1997

15–24 May	STS-84/Atlantis
1–17 July	STS-94/Columbia
5 August	Soyuz-TM 26/Mir PE-24 launch
7–19 August	STS-85/Discovery

22 AUGUST

1997 EVA 7	**Duration:** 5:30
World EVA 155	**Spacecraft/mission:** Mir PE-24
Russian EVA 79	**Crew:** Anatoli Solovyov, Pavel Vinogradov, Michael Foale (NASA)
Space Station EVA 88	**Spacewalkers:** Anatoli Solovyov, Pavel Vinogradov
Mir EVA 62	**Purpose:** Assess damage from Progress collision; connect solar array power cables inside Spektr; inspect Spektr interior for leaks; retrieve science and personal items from Spektr

On 25 June, during an attempt to dock a remote-controlled Progress resupply ship, a collision occurred with Mir's Spektr module. The accident caused a puncture in Spektr that led to a drop in cabin pressure in the entire station. Mir then started to spin at a rate of once every 6 minutes. The crew was able to close Spektr's hatch and stabilize the station, avoiding an emergency evacuation. The collision raised concerns regarding the safety of Mir and future Shuttle-Mir missions. An internal NASA evaluation and two independent assessments took place before NASA Administrator Daniel Goldin approved the succeeding Shuttle-Mir missions. This on-orbit accident was the first spacecraft depressurization caused by an accident in the history of human spaceflight. Prior to this EVA, there was concern about the large amount of debris and loose cables in the depressurized Spektr module that could be hazardous to the spacewalkers. Anatoli Solovyov and Pavel Vinogradov were the first to begin repairing the damage caused by the collision. They reconnected Spektr's power cables to a new custom hatch, surveyed Spektr's depressurized interior, and recovered Michael Foale's personal and experimental materials.[1]

---

1. "Mission Archives: STS-86", Jeanne Ryba, editor, accessed 2 November 2015, (*http://www.nasa.gov/mission_pages/shuttle/shuttlemissions/archives/sts-86.html*).

FIGURE 1. **Damaged Spektr Solar Array.** A 70-mm view of Russia's Mir space station backdropped against a cloud-covered Earth was photographed during a fly-around by the Space Shuttle Atlantis following the conclusion of joint docking activities between the Mir-24 and STS-86 crews. One of the solar array panels on the Spektr module shows damage incurred during the impact of a Russian uncrewed Progress resupply ship that collided with the space station on 25 June 1997. (NASA STS086-710-007)

6 SEPTEMBER

1997 EVA 8	**Duration:** 6:00
World EVA 156	**Spacecraft/mission:** Mir PE-24
Russian EVA 80/ U.S. EVA 78	**Crew:** Anatoli Solovyov, Pavel Vinogradov, Michael Foale (NASA)
Space Station EVA 89	**Spacewalkers:** Anatoli Solovyov, Michael Foale
Mir EVA 63	**Purpose:** Inspect Spectr exterior for air leaks; rotate two Spektr solar arrays

Foale accompanied Solovyov on a 6-hour spacewalk to inspect the Spektr module and adjust its solar panels. The EVA was particularly dangerous due to the risk created by Spektr's debris and sharp edges resulting from the damaged structure that could penetrate the spacesuits. Solovyov performed most of the detailed work. He removed layers of thermal blankets from Spektr with a sharp tool. Foale later stated during an oral history interview that Solovyov "was using a razor [sic] knife to basically cut away at the insulation. We had a camera with us called the Gleesa. It showed the hull to be damaged in that area though the exterior panels were buckled and bent there and some of the support was bent." The source of the leak on Spektr was never found. Meanwhile, Foale shot video footage and operated the Strela, a long cargo boom, used to maneuver Solovyov. The spacewalkers also adjusted a solar panel, increasing Mir's power supply by 10 percent. Though the EVA was successfully completed, concern over the safety of Mir and Shuttle-Mir missions increased when Solovyov discussed his frustration and breathing problems during the EVA. Ground Chief Vladimir Solovyov acknowledged he may have "over-assigned" Foale and Solovyov's already dangerous tasks.[2]

25 September STS-86/Atlantis launch

1 OCTOBER

1997 EVA 9	**Duration:** 5:01
World EVA 157	**Spacecraft/mission:** STS-86
Russian EVA 81/U.S. EVA 79	**Crew:** James Wetherbee, Michael Bloomfield, Scott Parazynski, Wendy Lawrence, David Wolf, Jean-Loup Chretien (CNES), Vladimir Titov (Russian Space Agency)
Shuttle EVA 38	**Spacewalkers:** Scott Parazynski, Vladimir Titov
	Purpose: Retrieve four Mir Environmental Effects Payloads (MEEPS) materials; attach solar array cap to docking module; test the Simplified Aid for EVA Rescue (SAFER) jet packs; test U.S. and Russian tethers

Vladimir Titov and Scott Parazynski's spacewalk marked the first joint U.S.-Russian EVA with a Shuttle docked to the Mir. Their 5-hour, 1-minute spacewalk required them to attach a 121-pound (54.88-kilogram) solar array cap to Mir's docking module, allowing crewmembers to seal off Spektr's suspected air leak. The astronauts also recovered four MEEPS. A failed valve prevented the SAFER jet packs from firing, causing the test to be declared unsuccessful. The jet packs were designed to assist spacewalkers maneuver back to Mir or a future space station if they became untethered.[3]

6 October STS-86/Atlantis landing

---

2. "Mission Archives: STS-86," Jeanne Ryba, editor, accessed 2 November 2015, (*http://www.nasa.gov/mission_pages/shuttle/shuttlemissions/archives/sts-86.html*); "Possible Source of Leak on Mir Found," CNN, accessed 2 November 2015, (*http://www.cnn.com/TECH/9709/06/mir.update*); "Shuttle-Mir Stories–Foale on his EVA," Kim Dismukes, editor, (*http://spaceflight.nasa.gov/history/shuttle-mir/history/h-flights.htm*); "U.S. Mir Residents: Michael Foale," Kim Dismukes, editor, (*http://spaceflight.nasa.gov/history/shuttle-mir/people/p-r-foale.htm*).

3. "Mission Archives: STS-86," Jeanne Ryba, editor, accessed October 20, 2014, (*http://www.nasa.gov/mission_pages/shuttle/shuttlemissions/archives/sts-86.html*).

20 OCTOBER

1997 EVA 10	**Duration:** 6:10
World EVA 158	**Spacecraft/mission:** Mir PE-24
Russian EVA 82	**Crew:** Anatoli Solovyov, Pavel Vinogradov, David Wolf (NASA)
Space Station EVA 90	**Spacewalkers:** Anatoli Solovyov, Pavel Vinogradov
Mir EVA 64	**Purpose:** Connect two of Spektr's solar array orientation cables

Solovyov and Vinogradov connected two of Spektr's three cables, allowing the solar arrays to reorient and maximize solar efficiency. The solar arrays provided electrical power for the Spektr, which was now depressurized and unusable after the collision.[4]

3 NOVEMBER

1997 EVA 10	**Duration:** 6:05
World EVA 159	**Spacecraft/mission:** Mir PE-24
Russian EVA 83	**Crew:** Anatoli Solovyov, Pavel Vinogradov, David Wolf (NASA)
Space Station EVA 91	**Spacewalkers:** Anatoli Solovyov, Pavel Vinogradov
Mir EVA 65	**Purpose:** Transfer solar array from Kvant 1 to base block; install Vozduk CO_2 system; retrieve Kvant 2 science data; release Sputnik anniversary satellite

The spacewalkers detached the solar array from Kvant 1, then folded and transferred it to the module's base block. They retrieved data from Kvant 2's science experiments for further study. The Vozduk CO_2, a carbon dioxide removal system, was also installed. The cosmonauts then celebrated the anniversary of Earth's first artificial satellite by launching a Sputnik mock-up. Upon reentering the spacecraft, the Kvant 2 outer hatch would not hermetically seal due to the "C" clamp's deterioration. The inner compartment was used as a backup airlock.[5]

6 NOVEMBER

1997 EVA 12	**Duration:** 6:19
World EVA 160	**Spacecraft/mission:** Mir PE-24 **Crew:** Anatoli Solovyov, Pavel Vinogradov, David Wolf (NASA)
Russian EVA 84	**Spacewalkers:** Anatoli Solovyov, Pavel Vinogradov
Space Station EVA 92	**Purpose:** Install new solar array to Kvant 1; install Vozduk CO_2 outlet diffuser; install backup bolts on Kvant 2
Mir EVA 66	

4. *Russian Spacesuits,* Isaak P. Abramov and Å. Ingemar Skoog, Springer, 2003, p. 316.

5. *Praxis Manned Spaceflight Log: 1961–2006*, Tim Furniss and David J. Shayler with Michael D. Shayler, Springer, 2007, p. 606; Russian Spacesuits, Isaak P. Abramov and Å. Ingemar Skoog, Springer, 2003, p. 316.

The cosmonauts completed the goals of the previous spacewalk and replaced the solar array attached to Kvant 1. For the first time, the Strela cranes handed off a stowed solar array. Since the Kvant 2 hatch could not repressurize correctly during the previous excursion, the spacewalkers inspected the hatch and fastened it with additional bolts.[6]

19 November STS-87/Columbia launch

24 NOVEMBER

1997 EVA 13	**Duration:** 7:43
World EVA 161	**Spacecraft/mission:** STS-87
Japanese EVA 1	**Crew:** Kevin Kregel, Steven Lindsey, Winston Scott, Kalpana Chawla, Takao Doi (NASDA), Leonid Kadenyuk (NSAU)
U.S. EVA 80	**Spacewalkers:** Winston Scott, Takao Doi
Shuttle EVA 39	**Purpose:** Unscheduled EVA to manually capture and berth SPARTAN 201; demonstrate ISS end-to-end EVA assembly and maintenance operations in time remaining

NASA delayed SPARTAN 201's scheduled release into orbit by a day, allowing its accompanying spacecraft, the Solar and Heliospheric Observatory (SOHO), to return online. On November 22, Kalpana Chawla operated the Remote Manipulator System (RMS), Columbia's mechanical arm, to unberth and then release SPARTAN. However, a problem with the attitude control system, which allowed precise pointing toward solar targets, prevented the spacecraft from completing a pirouette turn successfully. Chawla regrasped and released SPARTAN, causing it to spin approximately two degrees per second. Kevin Kregel attempted to match Columbia's rotation with SPARTAN's, but the maneuver was cancelled by the flight director. Only one EVA was originally planned for the mission, but two days later, Winston Scott and Takao Doi executed an unscheduled EVA to retrieve the free-floating SPARTAN. Doi became the first Japanese astronaut to perform a spacewalk. After recapturing the satellite, they resumed preparations for the on-orbit assembly of the International Space Station (ISS). The spacewalkers' work focused on testing tools and equipment and demonstrating EVA operations needed for ISS assembly and maintenance.[7]

---

6. *Russian Spacesuits,* Isaak P. Abramov and Å. Ingemar Skoog, Springer, 2003, p. 316.

7. *NASA Science: Missions: Spartan 201*, accessed 2 November 2015, (*http://science.nasa.gov/missions/spartan*); "Mission Archives: STS-87," Jeanne Ryba, editor, accessed 2 November 2015, (*http://www.nasa.gov/mission_pages/shuttle/shuttlemissions/archives/sts-87.html*); "Missions: SPARTAN 201," National Aeronautics and Space Administration, accessed 2 November 2015, (*https://archive.org/details/NIX-KSC-97PC-1460*); *Solar and Heliospheric Observatory*, National Aeronautics and Space Administration, accessed 2 November 2015, (*http://www.nasa.gov/mission_pages/soho/index.html*).

3 DECEMBER

1997 EVA 14	**Duration:** 4:59
World EVA 162	**Spacecraft/mission:** STS-87
Japanese EVA 2	**Crew:** Kevin R. Kregel, Steven Lindsey, Winston Scott, Kalpana Chawla, Takao Doi (JAXA), Leonid Kadenyuk (NSAU)
U.S. EVA 81	**Spacewalkers:** Winston Scott, Takao Doi
Shuttle EVA 40	**Purpose:** Demonstrate ISS end-to-end EVA assembly and maintenance operations; deploy and retrieve Autonomous Extravehicular Activity Robotic Camera (AERCam) Sprint video camera

Scott and Doi's first unscheduled spacewalk had the objective of retrieving and berthing the SPARTAN satellite manually, leaving their initial goals of the originally scheduled EVA to demonstrate ISS end-to-end EVA assembly and maintenance operations using the ISS crane incomplete. The second EVA allowed the astronauts to finish testing the EVA crane and the AERCam Sprint. The spacewalkers manually operated the crane, which allowed future ISS EVA crews to move Orbital Replacement Units (ORUs), tools and equipment during ISS EVA assembly and maintenance operations efficiently. Meanwhile, IVA crewmember and Pilot Steven Lindsey maneuvered the AERCam Sprint remotely and demonstrated its benefit as a free-flying, extra set of "eyes" for future Shuttle orbiter and ISS inspections. The AERCam Sprint was a small soccer ball size spherical free-flying spacecraft with a soft impact-absorbing outer layer, 12 non-contaminating nitrogen gas thrusters and two television cameras. After a 30-minute flight, AERCam Sprint was manually retrieved and returned to the orbiter's airlock.[8]

5 December STS-87/Columbia landing

8. "AERCam Sprint," Kim Dismukes, editor, accessed 18 December 2013, (*http://spaceflight.nasa.gov/station/assembly/sprint*); "STS-87 Day 15 Highlights," National Aeronautics and Space Administration, accessed 17 December 2013, (*http://science.ksc.nasa.gov/shuttle/missions/sts-87/sts-87-day-15-highlights.html*).

1998

9 JANUARY

1998 EVA 1	**Duration:** 3:06
World EVA 163	**Spacecraft/mission:** Mir PE-24
Russian EVA 85	**Crew:** Anatoli Solovyov, Pavel Vinogradov, David Wolf (NASA)
Space Station 93	**Spacewalkers:** Anatoli Solovyov, Pavel Vinogradov
Mir EVA 67	**Purpose:** Inspect and film the damaged Kvant 2 airlock hatch, retrieve Optical Properties Monitor (OPM) experiment from docking module

The OPM was contributed by the NASA Glenn Research Center to study optical and thermal properties in space further. The cosmonauts retrieved the OPM and prepared for its return to Earth. The Kvant 2 hatch, which had been damaged and repaired using the Strela boom, was inspected further and filmed for documentation. Repairs were then made to the hatch to complete the EVA.[1]

14 JANUARY

1998 EVA 2	**Duration:** 3:52
World EVA 164	**Spacecraft/mission:** Mir PE-24
Russian EVA 86/ U.S. EVA 82	**Crew:** Anatoli Solovyov, Pavel Vinogradov, David Wolf (NASA)
	Spacewalkers: Anatoli Solovyov, David Wolf
Space Station EVA 94	**Purpose:** Use the Space Portable Spectroreflectormeter (SPSR) experiment on Kvant 2 radiator; Kvant 2 airlock hatch inspected and bolted closed
Mir EVA 68	

This EVA performed by David Wolf and Anatoli Solovyov was the third joint U.S.-Russian spacewalk. Wolf operated the SPSR and inspected Mir's exterior surfaces. By directing the handheld SPSR toward Mir, the spacewalkers accumulated exceptional data from the station's external surfaces. SPSR measured the absorption proficiencies of solar equipment, reflective mirrors, and other orbiting surfaces. Solovyov made additional repairs to the Kvant 2 hatch. The Kvant 2 module included an airlock unit, used for furthering EVA capabilities, biological, and Earth research.[2]

---

1. "Glenn's Participation in Shuttle-Mir Missions," Kathleen Zona, editor, accessed 13 February 2014, (*http://www.nasa.gov/centers/glenn/about/history/mirlewis.html*); *Russian Spacesuits,* Isaak P. Abramov and Å. Ingemar Skoog, Springer, 2003, p. 316.

2. "Biographical Data: David A. Wolf," National Aeronautics and Space Agency, accessed 28 December 2013, (*http://www.jsc.nasa.gov/Bios/htmlbios/wolf.html*); "Kvant-2 Module," Kim Dismukes, editor, accessed 3 January 2014, (*http://spaceflight.nasa.gov/history/shuttle-mir/spacecraft/s-mir-kvant2-main.htm*); *Phase 1 Program Joint Report*, George C. Nield and Pavel Mikhailovich Vorobiev, editors, accessed 8 January 2014, (*http://spaceflight.nasa.gov/spacenews/factsheets/pdfs/issphase1.pdf*); "Section 8–Extravehicular Activity," Aleksandr Pavlovich Aleksandrov and Richard Fullerton, accessed 2 January 2014, (*http://spaceflight.nasa.gov/history/shuttle-mir/references/documents/jr-sec8.pdf*); "In-Situ Materials Experiments on the Mir

22–31 January STS-89/Endeavour

29 January Soyuz-TM 27/Mir PE-25 launch

19 February Soyuz-TM 26/Mir PE-24 landing

1 APRIL

1998 EVA 3	**Duration:** 6:26
World EVA 165	**Spacecraft/mission:** Mir PE-25
Russian EVA 87	**Crew:** Talgat Musabayev, Nikolai Budarin, Leopold Eyherts (European Space Agency), Andrew Thomas (NASA)
Space Station EVA 95	
Mir EVA 69	**Spacewalkers:** Talgat Musabayev, Nikolai Budarin
	Purpose: Install handrails and foot restraints on Spektr in preparation for the repair of the damaged solar arrays

A scheduled spacewalk on 3 March was postponed to 1 April due to a broken wrench preventing the Kvant 2 airlock hatch from opening. Talgat Musabayev and Nikolai Budarin installed handrails and foot restraints on Spektr to prepare for a future mission to repair solar arrays.[3]

6 APRIL

1998 EVA 4	**Duration:** 4:23
World EVA 166	**Spacecraft/mission:** Mir PE-25
Russian EVA 88	**Crew:** Talgat Musabayev, Nikolai Budarin, Leopold Eyherts (European Space Agency), Andrew Thomas (NASA)
Space Station EVA 96	
Mir EVA 70	**Spacewalkers:** Talgat Musabayev, Nikolai Budarin
	Purpose: Reinforce Spektr solar array; retrieve Komza experiment; replace Portable Power Plant (PPP); replace thruster on Sofora boom

The spacewalkers reinforced the damaged Spektr solar array by installing a support brace. Installing the support brace required more time than planned, leaving the cosmonauts only an hour to replace the PPP and the thruster attached to the Sofora boom. They revised their plan and only began work on the PPP. Due to an error by Mission Control in commanding the Mir to drift, the EVA was ended early. This drift was misdiagnosed as a depletion of fuel of the VDU orientation engine when the EVA crew was ordered to end the EVA and return to the airlock.[4]

Station," Donald R. Wilkes and Melvin R. Carruth, accessed 8 December 2013, (*http://spaceflight.nasa.gov/history/shuttle-mir/science/iss/sc-iss-spsr.htm*).

3. *Russian Spacesuits,* Isaak P. Abramov and Å. Ingemar Skoog, Springer, 2003, p. 316.

4. *Russian Spacesuits,* Isaak P. Abramov and Å. Ingemar Skoog, Springer, 2003, p. 316; *Shuttle Mir: The United States and Russia Share History's Highest Stage*, Clay Morgan, accessed 2 February 2014, (*http://history.nasa.gov/SP-4225/documentation/mir-summaries/mir25/mr.htm*).

11 APRIL

1998 EVA 5	**Duration:** 6:25
World EVA 167	**Spacecraft/mission:** Mir PE-25
Russian EVA 89	**Crew:** Talgat Musabayev, Nikolai Budarin, Leopold Eyherts (European Space Agency), Andrew Thomas (NASA)
Space Station EVA 97	**Spacewalkers:** Talgat Musabayev, Nikolai Budarin
Mir EVA 71	**Purpose:** Remove and jettison the Vynosnaya Dvigatel'naya Ustanovka (VDU) roll thruster; recover experiment on Rapana structure and then remove and jettison the Rapana structure

Budarin and Musabayev completed work from the previous EVA. They discarded the PPP along with Sofora's boom jet assembly, known as the "VDU." The newer thruster provided better assistance in maneuvering Mir.[5]

17 April–3 May STS-90/Columbia

17 APRIL

1998 EVA 6	**Duration:** 6:33
World EVA 168	**Spacecraft/mission:** Mir PE-25
Russian EVA 90	**Crew:** Talgat Musabayev, Nikolai Budarin, Leopold Eyherts (European Space Agency), Andrew Thomas (NASA)
Space Station EVA 98	**Spacewalkers:** Talgat Musabayev, Nikolai Budarin
Mir EVA 72	**Purpose:** Prepare installation of new VDU to Sofora boom; remove and secure Ferma-3 and Rapana trusses to the Mir exterior

To prepare the installation of a new VDU thruster to the Sofora structure, the cosmonauts removed equipment that was no longer in use. With the Komza experiment still intact, Musabayev and Budarin dismantled and stored the Rapana for future use. Rapana is a segment of the VDU that secured mounted experiments. The pair also angled and bolted Sofora at 35 degrees to transfer the thruster during the next spacewalk more easily.[6]

---

5. Ibid.

6. Ibid.

22 APRIL

1998 EVA 7	**Duration:** 6:21
World EVA 169	**Spacecraft/mission:** Mir PE-25
Russian EVA 91	**Crew:** Talgat Musabayev, Nikolai Budarin, Leopold Eyherts (European Space Agency), Andrew Thomas (NASA)
Space Station EVA 99	**Spacewalkers:** Talgat Musabayev, Nikolai Budarin
Mir EVA 73	**Purpose:** Install new VDU on Sofora and erect to vertical position

The cosmonauts jettisoned the remainder of the old, exhausted VDU thruster into space at an angle below and in front of Mir, allowing the boom jet to burn up in Earth's atmosphere within a year. They also installed the new PPP onto Sofora and repositioned the boom vertically. Meanwhile, Andrew Thomas documented the cosmonauts' progress with video and photography. Because Musabayev lost telemetry capabilities, Thomas provided system data to ground control.[7]

2–12 June STS-91/Discovery

13 August Soyuz-TM 28/Mir PE-26 launch

25 August Soyuz-TM 27/Mir PE-25 landing

15 SEPTEMBER

1998 EVA 8	**Duration:** 0:30
World EVA 170	**Spacecraft/mission:** Mir PE-26
Russian EVA 92	**Crew:** Gennady Padalka, Sergei Avdeyev
Space Station EVA 100	**Spacewalkers:** Gennady Padalka, Sergei Avdeyev
Mir EVA 74	**Purpose:** Connect Spektr solar array cable

Sergei Avdeyev and Gennady Padalka performed an EVA inside the damaged Spektr. While inside the module, they connected the solar array orientation cable. The spacewalkers mated Spektr electrical connectors to the solar array servomotors.[8]

29 October–7 November STS-95/Discovery

7. Ibid.

8. *Russian Spacesuits,* Isaak P. Abramov and Å. Ingemar Skoog, Springer, 2003, p. 316.

10 NOVEMBER

1998 EVA 9	**Duration:** 5:50
World EVA 171	**Spacecraft/mission:** Mir PE-26
Russian EVA 93	**Crew:** Gennady Padalka, Sergei Avdeyev
Space Station EVA 101	**Spacewalkers:** Gennady Padalka, Sergei Avdeyev
Mir EVA 75	**Purpose:** Manually launch Sputnik 41 demo satellite; deploy, install, and retrieve various French experiments

Padalka and Avdeyev installed French scientific equipment: an experimental solar array on the docking module and a Dvikon material exposure on the Igla antenna. Furthermore, they retrieved samples of the thruster exhaust, as well as the Keramika, Danko, Solyaris, and SMMK experiments. The Komet, Spika, Migmas, and Sprut experiments were deployed on Kvant 2. To celebrate Sputnik's 41st anniversary launch, the cosmonauts manually launched a demo of the first human-made satellite to orbit Earth. [9]

4 December STS-88/Endeavour launch

7 DECEMBER

1998 EVA 10	**Duration:** 7:21
World EVA 172	**Spacecraft/mission:** STS-88
U.S. EVA 83	**Crew:** Robert Cabana, Frederick Sturckow, Nancy Currie, Jerry Ross, James Newman, Sergei Krikalev (Russian Space Agency)
Shuttle EVA 41	**Spacewalkers:** Jerry Ross, James Newman
	Purpose: Perform first EVA to assemble the ISS; first of three scheduled EVAs. Connect and activate all power cables between Zarya and Unity

During their 12-day mission, the STS-88 crew's objective was to power the U.S.-built Unity Node by mating it with the Russian-built Zarya. Prior to the EVA, the crew remotely captured the orbiting Zarya, commonly known as the Functional Cargo Block (FGB), and coupled it to Unity. Spacewalkers Jerry Ross and James Newman successfully connected about 40 power and data transmission cables. The astronauts then covered each connector with a thermal cover. Power from the FGB activated Unity's computers, avionics, and heaters. This was the first EVA for the ISS assembly.[10]

---

9. *Praxis Manned Spaceflight Log: 1961–2006*, Tim Furniss and David J. Shayler with Michael D. Shayler, Springer, 2007, p. 629; *Russian Spacesuits,* Isaak P. Abramov and Å. Ingemar Skoog, Springer, 2003, p. 316.

10. "International Space Station: Unity Node." Jerry Wright, editor, accessed 10 January 2013, (*http://www.nasa.gov/mission_pages/station/structure/elements/node1.html*); "Mission Archives: STS-88," Jeanne Ryba, editor, accessed 4 December 2013, (*http://www.nasa.gov/mission_pages/shuttle/shuttlemissions/archives/sts-88.html*); "STS-88 Day 5 Highlights," National Aeronautics and Space Administration, accessed 10 January 2013, (*http://science.ksc.nasa.gov/shuttle/missions/sts-88/sts-88-day-05-highlights.html*); "STS-88 Extravehicular Activities," Kim Dismukes, editor, accessed 4 December 2013, (*http://spaceflight.nasa.gov/shuttle/archives/sts-88/eva/index.html*).

9 DECEMBER

1998 EVA 11	**Duration:** 7:02
World EVA 173	**Spacecraft/mission:** STS-88
U.S. EVA 84	**Crew:** Robert Cabana, Frederick Sturckow, Nancy Currie, Jerry Ross, James Newman, Sergei Krikalev (Russian Space Agency)
Shuttle EVA 42	**Spacewalkers:** Jerry Ross, James Newman
	Purpose: Second of three scheduled EVAs. Install S-band Early Communication antennas; remove launch restraints; unfold Zarya antenna

The Endeavour EVA crewmembers installed two S-band Early Communication (ECOMM) antennas to Unity. Following the EVA, a corresponding avionics gear was installed inside the ISS, allowing the antennas to transmit a feed of the Station's monitors to U.S. flight controllers. Ross and Newman also removed launch restraints from Unity's four Common Berthing Mechanisms (CBMs). The four CBMs will then be able to dock future modules to the Station. The spacewalkers then mounted EVA handrails to Zarya's exterior. Since time remained in the EVA, Newman used a 10-foot-long (3.05-meter) grappling hook to unfurl one of Zarya's two jammed antennas.[11]

12 DECEMBER

1998 EVA 12	**Duration:** 6:59
World EVA 174	**Spacecraft/mission:** STS-88
U.S. EVA 85	**Crew:** Robert Cabana, Frederick Sturckow, Nancy Currie, Jerry Ross, James Newman, Sergei Krikalev (Russian Space Agency)
Shuttle EVA 43	**Spacewalkers:** Jerry Ross, James Newman
	Purpose: Third of three scheduled EVAs. Deploy two U.S. portable foot restraints and a tool stanchion; aid the release of two FGB TORU antennas; install FGB EVA handrails and reinstall FGB Komplast panel, and SAFER self-rescue demo

In this final EVA for the STS-88 crew, Ross and Newman tested EVA safety equipment and finalized preparations for ISS's future assembly. They first disconnected unnecessary cable ties from the previous EVA and the cables used to connect Unity and Zarya initially. In addition, they mounted EVA handrails on Zarya for future spacewalkers. A large EVA tool bag, which included portable foot restraints and EVA tools such as wrenches and ratchets, was fixed to the Station's exterior. The attached items later assisted many ISS spacewalkers. Both astronauts successfully tested the SAFER jet backpacks that failed during the STS-86 EVA. The jet backpacks can propel astronauts back to

11. "Mission Archives: STS-88," Jeanne Ryba, editor, accessed 14 December 2013, (*http://www.nasa.gov/mission_pages/shuttle/shuttlemissions/archives/sts-88.html*); "STS-88 Day 7 Highlights," National Aeronautics and Space Administration, accessed 13 December 2013, (*http://science.ksc.nasa.gov/shuttle/missions/sts-88/sts-88-day-07-highlights.html*); "STS-88 Extravehicular Activities," Kim Dismukes, editor, accessed 4 December 2013, (*http://spaceflight.nasa.gov/shuttle/archives/sts-88/eva/index.html*).

the Station if they become untethered. They also produced a thorough photographic assessment of the entire Space Station.[12]

15 December STS-88/Endeavour landing

12. "*Mission Highlights STS-88*," Johnson Space Center, accessed 2 January 2014, (*http://spaceflight.nasa.gov/spacenews/factsheets/pdfs/hilit88.pdf*); "STS-88 Day 10 Highlights," National Aeronautics and Space Administration, accessed 2 January 2014, (*http://science.ksc.nasa.gov/shuttle/missions/sts-88/sts-88-day-10-highlights.html*).

1999

8 February Soyuz-TM 28/Mir PE-26 landing

20 February Soyuz-TM 29/Mir PE-27 launch

16 APRIL

1999 EVA 1	**Duration:** 6:19
World EVA 175	**Spacecraft/mission:** Mir PE-27
Russian EVA 94	**Crew:** Victor Afanaseyev, Sergei Avdeyev, Jean-Pierre Haignere (European Space Agency)
Space Station EVA 102	**Spacewalkers:** Victor Afanasyev, Jean-Pierre Haignere
Mir EVA 76	**Purpose:** Test leak repair tool; deploy French experiments; launch amateur radio satellite Sputnik 99

Victor Afanasyev and Jean-Pierre Haignere were scheduled to test a new sealant device to repair small leaks. The test of the tool on a simulated hole in Kvant was cancelled when the hatch failed to open. The plan to deploy new organic detectors was also cancelled due to lack of time. The spacewalkers were able, however, to retrieve the French Micrometeoroid and Orbital Debris (MMOD) experiment from the station's exterior and run an experiment designed by French high schools with the help of the French space agency, Centre National d'Etudes Spatiales (CNES). A manual launch of the third and final French built amateur radio satellite, Sputnik 99, was also successful. This small satellite was called Sputnik 99 because it was launched in 1999.[1]

27 May STS-96/Discovery launch

29 MAY

1999 EVA 2	**Duration:** 7:55
World EVA 176	**Spacecraft/mission:** STS-96
U.S. EVA 86	**Crew:** Kent Rominger, Rick Husband, Ellen Ochoa, Tamara Jernigan, Daniel Barry, Julie Payette (Canadian Space Agency), Valeri Ivanovich Tokarev (Russian Space Agency)
Shuttle EVA 44	**Spacewalkers:** Tamara Jernigan, Daniel Barry
	Purpose: Transfer and externally install U.S. and Russian cargo cranes, two common foot restraints, and three EVA tool bags

1. *Praxis Manned Spaceflight Log: 1961–2006*, Tim Furniss and David J. Shayler with Michael D. Shayler, Springer, 2007, p. 629; *Russian Spacesuits,* Isaak P. Abramov and Å. Ingemar Skoog, Springer, 2003, p. 316.

During the second longest spacewalk thus far, Tamara Jernigan and Daniel Barry transferred two cargo cranes from Discovery's payload bay and mounted them to the ISS. Both devices, the Russian-built Strela crane and the U.S.-built ORU Transfer Device, also known as the EVA crane, are designed to move large components and crewmembers during ISS construction and maintenance. The spacewalkers also attached three bags of EVA tools and handrails for future construction of the Station. They then installed two common foot restraints that fit both U.S. and Russian spacesuit boots. Jernigan and Barry photographed and documented the Shuttle's painted surfaces, fitted an insulating cover over Unity's trunnion pin, and examined one of Unity's ECOMM antennas.[2]

6 June STS-96/Discovery landing

23–27 July STS-93/Columbia

23 JULY

1999 EVA 3	**Duration:** 6:07
World EVA 177	**Spacecraft/mission:** Mir PE-27
Russian EVA 95	**Crew:** Victor Afanaseyev, Sergei Avdeyev, Jean-Pierre Haignere (European Space Agency)
Space Station EVA 103	**Spacewalkers:** Victor Afanaseyev, Sergei Avdeyev
Mir EVA 77	**Purpose:** Install experimental Georgian-Russian communications antenna; retrieve French experiments Exobiology and Dvikon

Afanaseyev and Avdeyev installed the experimental Georgian-Russian elliptical reflector antenna on the Sofora truss to test a new generation of communications satellites. This experimental antenna was 6.4 meters in diameter and 1.1 meters in depth. Unfortunately, the experimental antenna failed to deploy completely despite Afanaseyev and Avdeyev's attempts to open it. They were only able to open it about 80 to 90 percent. The spacewalkers were also unable to locate the source of a small leak in Kvant 2. They successfully retrieved samples, however, from the French experiments Exobiology and Dvikon. Towards the end of the EVA, Afanaseyev's spacesuit had a thermoregulation failure.[3]

---

2. "Mission Archives: STS-96," Jeanne Ryba, editor, accessed 8 December 2013, (*http://www.nasa.gov/mission_pages/shuttle/shuttlemissions/archives/sts-96.html*); "STS-Extravehicular Activities," Kim Dismukes, editor, accessed 8 December 2013, (*http://spaceflight.nasa.gov/shuttle/archives/sts-96/eva/index.html*).

3. *Praxis Manned Spaceflight Log: 1961–2006*, Tim Furniss and David J. Shayler with Michael D. Shayler, Springer, 2007, p. 638; *Russian Spacesuits,* Isaak P. Abramov and Å. Ingemar Skoog, Springer, 2003, p. 316.

28 JULY

1999 EVA 4	**Duration:** 5:22
World EVA 178	**Spacecraft/mission:** Mir PE-27
Russian EVA 96	**Crew:** Victor Afanaseyev, Sergei Avdeyev, Jean-Pierre Haignere (European Space Agency)
Space Station EVA 104	**Spacewalkers:** Victor Afanaseyev, Sergei Avdeyev
Mir EVA 78	**Purpose:** Fully deploy and jettison experimental Georgian-Russian communications antenna; install experiments and retrieve samples

During this EVA, the cosmonauts successfully deployed the experimental communications antenna from the previous spacewalk. After the completely deployed communications antenna was evaluated, the cosmonauts manually disconnected the antenna and jettisoned it away from Mir. The Indicator and Sprut-4 experiments were installed on the exterior, and the Danko-M and Ekran-D experiments were successfully retrieved. The spacewalkers also replaced the cassettes of the Migmas ion spectrometer.[4]

28 August Soyuz-TM 29/Mir PE-27 landing

19 December STS-103/Discovery launch

22 DECEMBER

1999 EVA 5	**Duration:** 8:15
World EVA 179	**Spacecraft/mission:** STS-103
U.S. EVA 87	**Crew:** Curtis Brown, Scott Kelly, Steven Smith, Michael Foale, John Grunsfeld, Claude Nicollier (European Space Agency), Jean-François Clervoy (European Space Agency)
Shuttle EVA 45	**Spacewalkers:** Steven Smith, John Grunsfeld
	Purpose: Hubble Servicing Mission 3A, the first of three scheduled EVAs

This EVA was the first of three made over the period of three days in order to service and repair the decade-old Hubble Space Telescope (HST). Steven Smith and John Grunsfeld were the first pair of astronauts to perform these services. They began by replacing HST's three Rate Sensor Units (RSUs), each of which contained two gyroscopes. They then purged the coolant located in valves on the Near Infrared Camera and Multi-Object Spectrometer (NICMOS). This was done to prepare these instruments for servicing on the next Hubble Servicing Mission. The pair also installed six Voltage/Temperature Improvement Kits between the telescope's solar panels for each of its six batteries. The kits were designed to prevent the batteries, which had been in the telescope since its April 24, 1990, launch on STS-31, from overcharging and overheating. During the EVA, Smith and Grunsfeld

4. *Praxis Manned Spaceflight Log: 1961–2006*, Tim Furniss and David J. Shayler with Michael D. Shayler, Springer, 2007, p. 638; *Russian Spacesuits,* Isaak P. Abramov and Å. Ingemar Skoog, Springer, 2003, p. 316.

intended to take several close-up photographs of the newly installed kits, but had unforeseen difficulty removing one of the old RSUs and opening the valves and removing caps to purge coolant from the Near Infrared Camera and Multi-Object Spectrometer. This difficulty caused the RSU replacement to take much longer than planned, and by its end, the EVA clocked in at 8 hours, 15 minutes, making it the second longest spacewalk in history thus far.[5]

23 DECEMBER

1999 EVA 6	**Duration:** 8:10
World EVA 180	**Spacecraft/mission:** STS-103
ESA EVA 1/U.S. EVA 88	**Crew:** Curtis Brown, Scott Kelly, Steven Smith, Michael Foale, John Grunsfeld, Claude Nicollier (European Space Agency), Jean-François Clervoy (European Space Agency)
Shuttle EVA 46	**Spacewalkers:** Michael Foale, Claude Nicollier
	Purpose: Hubble Servicing Mission 3A, the second of three scheduled EVAs

Michael Foale and Claude Nicollier were the second pair to perform maintenance on HST on the second day of its three-consecutive-day servicing. Foale and Nicollier replaced HST's computer with an operating system that would run 20 times faster and contain 6 times the memory. They also replaced one of the telescope's three 550-pound (249.48-kilogram) Fine Guidance Sensors. At only 5 minutes less than the previous day's excursion, this EVA became the third longest in history thus far.[6]

24 DECEMBER

1999 EVA 7	**Duration:** 8:08
World EVA 181/ U.S. EVA 89	**Spacecraft/mission:** STS-103
Shuttle EVA 47	**Crew:** Curtis Brown, Scott Kelly, Steven Smith, Michael Foale, John Grunsfeld, Claude Nicollier (European Space Agency), Jean-François Clervoy (European Space Agency)
	Spacewalkers: Steven Smith, John Grunsfeld
	Purpose: Hubble Servicing Mission 3A, the third of three scheduled EVAs

The EVA team of Smith and Grunsfeld performed the third and final day's work on HST. They replaced one of Hubble's radio transmitters that sent scientific data from the telescope to Earth and

5. "Mission Archives: STS-103," Jeanne Ryba, editor, accessed 6 January 2014, (*http://www.nasa.gov/mission_pages/shuttle/shuttlemissions/archives/sts-103.html*); "STS-103 Day 3 Highlights," National Aeronautics and Space Administration, accessed 6 January 2014, (*http://science.ksc.nasa.gov/shuttle/missions/sts-103/sts-103-day-03-highlights.html*); *STS-103, Mission Control Center: Status Report #06*, NASA Johnson Space Center, accessed 6 January 2014, (*http://spaceflight.nasa.gov/spacenews/reports/sts103/STS-103-06.html*).

6. "Mission Archives: STS-103," Jeanne Ryba, editor, accessed 6 January 2014, (*http://www.nasa.gov/mission_pages/shuttle/shuttlemissions/archives/sts-103.html*); "STS-103 Day 4 Highlights," National Aeronautics and Space Administration, accessed 6 January 2014, (*http://science.ksc.nasa.gov/shuttle/missions/sts-103/sts-103-day-04-highlights.html*).

FIGURE 2. Hubble STS-103 Payload Commander Steven Smith prepares to use a 35 mm camera during the final spacewalk of the STS-103 mission. Smith is standing on a foot restraint connected to the end of Discovery's Remote Manipulator System (RMS) robot arm. Astronaut John Grunsfeld translates along a handrail system on the Hubble Space Telescope in the background. (NASA STS103-E-5347)

ceased operating in 1998. As radio transmitters are normally very reliable, it was not expected that one would require replacement in orbit. The astronauts used newly created EVA tools designed specifically for the job. Smith and Grunsfeld also replaced the telescope's mechanical reel-to-reel recorder with a Digital Solid State Recorder, which had more than 10 times the previous storage capacity. The spacewalk was estimated to last roughly 7 hours, but like the previous two days' EVAs, it lasted more than 8 hours, partly due to difficulty hooking Grunsfeld's suit up to orbiter power in Discovery's airlock[7]

27 December STS-103/Discovery landing

7. "Mission Archives: STS-103," Jeanne Ryba, editor, accessed 6 January 2014, (*http://www.nasa.gov/mission_pages/shuttle/shuttlemissions/archives/sts-103.html*); "STS-103 Day 5 Highlights," National Aeronautics and Space Administration, accessed 6 January 2014, (*http://science.ksc.nasa.gov/shuttle/missions/sts-103/sts-103-day-05-highlights.html*).

2000

11–22 February STS-99/Endeavour

4 April Soyuz-TM 30/Mir PE-28 launch

12 MAY

2000 EVA 1	**Duration:** 5:30
World EVA 182	**Spacecraft/mission:** Mir PE-28
Russian EVA 97	**Crew:** Sergei Zalyotin, Alexander Kaleri
Space Station EVA 105	**Spacewalkers:** Sergei Zalyotin, Alexander Kaleri
Mir EVA 79	**Purpose:** Inspect and photograph short circuit damaged solar array cable on Kvant-1; test module leak repair sealant; retrieve a solar array test sample from docking module; conduct general inspection of Mir exterior

Cosmonauts Sergei Zalyotin and Alexander Kaleri performed the last Mir spacewalk. They completed the last general assessment, detailing Mir's condition and panoramically photographing its exterior. Various data samples were retrieved before the experiments were dismantled and jettisoned. An examination of a solar cell cable that previously short-circuited was also completed. Finally, the duo completed a delayed spacewalking task: the crew successfully tested a leak repair sealant on the station's hull.[1]

19 May STS-101/Atlantis launch

21 MAY

2000 EVA 2	**Duration:** 6:44
World EVA 183	**Spacecraft/mission:** STS-101
U.S. EVA 90	**Crew:** James Halsell Jr., Scott "Doc" Horowitz, Mary Ellen Weber, Jeffrey Williams, James Voss, Susan Helms, Yury Usachev (Russian Space Agency)
Shuttle EVA 48	**Spacewalkers:** Jeffrey Williams, James Voss
	Purpose: Re-secure and reorient U.S. and Russian cranes

The crew focused on transferring items from Atlantis to the ISS and inspecting the Orbital Replacement Unit Transfer Device and the Russian-built Strela. The two cranes were on the ISS for EVA maintenance. Unity's two ECOMM antennas, which allow U.S. flight controllers to monitor the Station,

---

1. *Praxis Manned Spaceflight Log: 1961–2006*, Tim Furniss and David J. Shayler with Michael D. Shayler, Springer, 2007, p. 653; *Russian Spacesuits,* Isaak P. Abramov and Å. Ingemar Skoog, Springer, 2003, p. 316.

were experiencing problems. The astronauts successfully replaced the antennas. They then installed eight additional EVA handrails for future spacewalks.[2]

29 May STS-101/Atlantis landing

16 June Soyuz-TM 30/Mir PE-28 landing

8 September STS-106/Atlantis launch

11 SEPTEMBER

2000 EVA 3
World EVA 184
Russian EVA 98/ U.S. EVA 91
Shuttle EVA 49

Duration: 6:14

Spacecraft/mission: STS-106

Crew: Terrence Wilcutt, Scott Altman, Daniel Burbank, Edward Lu, Richard Mastracchio, Yuri Malenchenko (Russian Space Agency), Boris Morokov (Russian Space Agency)

Spacewalkers: Edward Lu, Yuri Malenchenko

Purpose: Connect power, data, and communication cables between Zvezda and Zarya; install magnetometer backup navigation system

During this joint Russian and American spacewalk, Zvezda and Zarya's nine power cables were connected. Though both modules were Russian-made, Zvezda was the first solely Russian contribution—Boeing helped build Zarya. The Zvezda Service Module allowed the first human habitation of the ISS. The 14 by 13.5-foot (4.26 by 4.11-meter) module includes living quarters, life support systems, power distribution, propulsion, and command capabilities from ground flight controllers. Then, while constructing the backup navigation system, Edward Lu and Yuri Malenchenko were tethered 110 feet (33.53 meters) above the payload bay—the furthest a spacewalker traveled from the ISS at that point in operational history.[3]

20 September STS-106/Atlantis landing

11 October STS-92/Discovery launch

2. "Mission Archives: STS-101," Jeanne Ryba, editor, accessed 4 January 2014, (*http://www.nasa.gov/mission_pages/shuttle/shuttlemissions/archives/sts-101.html*); "STS-101 Day 3 Highlights," National Aeronautics and Space Administration, accessed 4 January 2014, (*http://science.ksc.nasa.gov/shuttle/missions/sts-101/sts-101-day-03-highlights.html*).

3. "International Space Station: Zvezda," Jerry Wright, editor, accessed 4 January 2014, (*http://www.nasa.gov/mission_pages/station/structure/elements/sm.html*); "STS-106 Day3 Highlights," National Aeronautics and Space Administration, accessed 4 January 2014, (*http://science.ksc.nasa.gov/shuttle/missions/sts-106/sts-106-day-03-highlights.html*); "STS-106 Extravehicular Activity," Kim Dismukes, editor, accessed 4 January 2014, (*http://spaceflight.nasa.gov/shuttle/archives/sts-106/eva/index.html*).

15 OCTOBER

2000 EVA 4	**Duration:** 6:28
World EVA 185	**Spacecraft/mission:** STS-92
U.S. EVA 92	**Crew:** Brian Duffy, Pamela Melroy, Leroy Chiao, Peter Wisoff, Michael Lopez-Alegria, William McArthur, Koichi Wakata (JAXA)
Shuttle EVA 50	**Spacewalkers:** Leroy Chiao, William McArthur
	Purpose: First of four scheduled EVAs. Relocate two S-band antennas to Zenith One (Z1) truss; connect power cables to Z1 truss; install safety tools for ISS assembly

Leroy Chiao and William McArthur continued assembling the ISS and installed additional safety tools for spacewalks. The spacewalkers relocated the S-band antennas from the Unity Node to the Z1 truss, to avoid the antennas' previous connection problems. Then, the astronauts installed 10 umbilical cables between the truss structure and Unity. The Z1 truss contains communication equipment and two plasma contractors to neutralize the ISS's static electrical charge.[4]

16 OCTOBER

2000 EVA 5	**Duration:** 7:07
World EVA 186	**Spacecraft/mission:** STS-92
U.S. EVA 93	**Crew:** Brian Duffy, Pamela Melroy, Leroy Chiao, Peter Wisoff, Michael Lopez-Alegria, William McArthur, Koichi Wakata (JAXA)
Shuttle EVA 51	**Spacewalkers:** Peter Wisoff, Michael Lopez-Alegria
	Purpose: Second of four scheduled EVAs. Mate Pressurized Mating Adapter (PMA) and Node 1 cables; prepare Z1 for solar array installation

IVA crewmember Koichi Wakata remotely installed the PMA-3 to the Unity's nadir port with the Shuttle's robotic arm. Spacewalkers Peter Wisoff and Michael Lopez-Alegria then connected the primary and secondary cables between the two components. They also prepared the Z1 truss for installation of U.S. solar arrays.[5]

---

4. "STS-92 Day 5 Highlights," National Aeronautics and Space Administration, accessed 5 January 2014, (*http://science.ksc.nasa.gov/shuttle/missions/sts-92/sts-92-day-05-highlights.html*); "STS-92 Extravehicular Activities," Kim Dismukes, editor, accessed 5 January 2014, (*http://spaceflight.nasa.gov/shuttle/archives/sts-92/eva/index.html*).

5. "Mission Archives: STS-92," Jeanne Ryba, editor, accessed 5 January 2014, (*http://www.nasa.gov/mission_pages/shuttle/shuttlemissions/archives/sts-92.html*); "STS-92 Extravehicular Activity," Kim Dismukes, editor, accessed 5 January 2014, (*http://spaceflight.nasa.gov/shuttle/archives/sts-92/eva/index.html*).

17 OCTOBER

2000 EVA 6	**Duration:** 6:48
World EVA 187	**Spacecraft/mission:** STS-92
U.S. EVA 94	**Crew:** Brian Duffy, Pamela Melroy, Leroy Chiao, Peter Wisoff, Michael Lopez-Alegria, William McArthur, Koichi Wakata (JAXA)
Shuttle EVA 52	**Spacewalkers:** Leroy Chiao and William McArthur
	Purpose: Third of four scheduled EVAs. Install two DC-to-DC converters to Z1 truss

Chiao and McArthur installed two DC-to-DC converters above the Z1 truss, enabling proper voltage conversion of the electricity generated by the Station's solar arrays. The system eliminates electricity and heat buildups that may damage equipment.[6]

18 OCTOBER

2000 EVA 7	**Duration:** 6:56
World EVA 188	**Spacecraft/mission:** STS-92
U.S. EVA 95	**Crew:** Brian Duffy, Pamela Melroy, Leroy Chiao, Peter Wisoff, Michael Lopez-Alegria, William McArthur, Koichi Wakata (JAXA)
Shuttle EVA 53	**Spacewalkers:** Peter Wisoff, Michael Lopez-Alegria
	Purpose: Fourth of four scheduled EVAs. Remove and stow Flight Releasable Grapple Fixture (FRGF) on Z1 Truss, deploy its utility tray; test SAFER hardware

During the final STS-92 spacewalk, Wisoff and Lopez-Alegria removed and stowed the FRGF on the Z1 Truss. They deployed the Z1 truss utility tray, which provides power to the United States lab. They then opened and cycled the Manual Berthing Mechanism for the first time. They also performed a successful protocol test of two SAFER jet backpacks, which allows spacewalkers to return to the Station if their tether unintentionally becomes detached.[7]

24 October STS-92/Discovery landing

31 October Soyuz-TM 31/ISS Expedition 1 launch

30 November STS-97/Endeavour launch

---

6. "NASA Glenn Contributions to the International Space Station (ISS) Electrical Power System," Kathleen Zona, editor, accessed 6 January 2014, (*http://www.nasa.gov/centers/glenn/about/fs06grc.html*); "STS-92 Extravehicular Activities," Kim Dismukes, editor, accessed 6 January 2014, (*http://spaceflight.nasa.gov/shuttle/archives/sts-92/eva/*).

7. "Mission Archives: STS-92," Jeanne Ryba, editor, accessed 7 January 2014, (*http://spaceflight.nasa.gov/shuttle/archives/sts-92/eva/index.html*);"STS-92 Extravehicular Activity," Kim Dismukes, editor, accessed 7 January 2014, (*http://spaceflight.nasa.gov/shuttle/archives/sts-92/eva/index.html*).

FIGURE 3. **Initial Assembly of ISS's Arrays.** Astronaut Joseph R. Tanner, mission specialist, waves to crewmembers inside Endeavour's cabin during an EVA to perform work on the ISS. The elbow of the RMS robotic arm and part of the newly deployed solar array panel are in the background. (NASA S97-E-5031)

3 DECEMBER

2000 EVA 8
World EVA 189
U.S. EVA 96
Shuttle EVA 54

Duration: 7:33

Spacecraft/mission: STS-97

Crew: Brent Jett, Jr., Michael Bloomfield, Joseph Tanner, Carlos Noriega, Marc Garneau (Canadian Space Agency)

Spacewalkers: Joseph Tanner, Carlos Noriega

Purpose: First of three scheduled EVAs. Attach P6 Integrated Truss to the Z1 Truss; install solar array on Unity's Z1 truss; deploy solar array

This was the first of three spacewalks performed by Joseph Tanner and Carlos Noriega for the assembly of the ISS. Marc Garneau moved the entire P6 truss with Endeavour's robotic arm into position on the Z1 truss structure, and Tanner and Noriega bolted it in place. Once the truss was secured, Garneau released it from the arm's grip and Michael Bloomfield used the arm to move Noriega around the array to connect nine power, command, and data cables. Meanwhile, Tanner readied the solar array's two blanket boxes to deploy the Space Station's starboard solar array. The pins holding the blanket boxes closed were to be released by computer command. Commander Brent Jett, Jr.'s first attempt to deploy the solar array was unsuccessful, and Tanner and Noriega stood by in case they needed to remove the pins manually, but Jett repeated the command and the starboard solar array was successfully deployed. Because of the problems with expanding the starboard solar array, flight controllers decided against releasing the pins of the other blanket box that would extend the portside solar array.[8]

5 DECEMBER

2000 EVA 9	**Duration:** 6:37
World EVA 190	**Spacecraft/mission:** STS-97
U.S. EVA 97	**Crew:** Brent Jett, Jr., Michael Bloomfield, Joseph Tanner, Carlos Noriega, Marc Garneau (Canadian Space Agency)
Shuttle EVA 55	**Spacewalkers:** Joseph Tanner and Carlos Noriega
	Purpose: Second of three scheduled EVAs. Install data and power cables to utilize energy generated by the solar arrays; move S-Band Assembly

Tanner and Noriega began their second spacewalk by surveying the newly erected solar array to analyze the condition of the tensioning system that extended one of its two solar array blankets. This survey and analysis allowed them to collect data and formulate a plan to further heighten the tension in the deployed array, in preparation for the following spacewalk. They also moved the S-band assembly to the top of the solar array tower. Finally, they removed the restraints that held a radiator against the side of the tower. The radiator would cool Destiny, the United States Laboratory Module, when it later launched to the ISS.[9]

7 DECEMBER

2000 EVA 10	**Duration:** 5:10
World EVA 191	**Spacecraft/mission:** STS-97
U.S. EVA 98	**Crew:** Brent Jett, Jr., Michael Bloomfield, Joseph Tanner, Carlos Noriega, Marc Garneau (Canadian Space Agency)
Shuttle EVA 56	**Spacewalkers:** Joseph Tanner, Carlos Noriega
	Purpose: Third of three scheduled EVAs. Adjust solar array tensioning cable; install the Floating Potential Probe (FPP) antennas on Node 1; install centerline camera cable on Node 1

---

8. "STS-97 Day 4 Highlights," National Aeronautics and Space Administration, accessed 11 January 2014, (*http://science.ksc.nasa.gov/shuttle/missions/sts-97/sts-97-day-04-highlights.html*); *STS-97, Mission Control Center: Status Report #07*, NASA Johnson Space Center, accessed 11 January 2014, (*http://spaceflight.nasa.gov/spacenews/reports/sts97/STS-97-07.html*).

9. "STS-97 Day 5 Highlights," National Aeronautics and Space Administration, accessed 11 January 2014, (*http://science.ksc.nasa.gov/shuttle/missions/sts-97/sts-97-day-05-highlights.html*); "STS-97 Day 6 Highlights," National Aeronautics and Space Administration, accessed 11 January 2014, (*http://science.ksc.nasa.gov/shuttle/missions/sts-97/sts-97-day-06-highlights.html*); *STS-97, Mission Control Center: Status Report #11*, NASA Johnson Space Center, accessed 11 January 2014, (*http://spaceflight.nasa.gov/spacenews/reports/sts97/STS-97-11.html*).

Tanner and Noriega started their final spacewalk of this mission with the impromptu task of furthering the pressure and rigidity of the starboard solar array. Crewmembers inside the Endeavour retracted the array by two to three feet (0.61 to 0.91 meters) for the spacewalkers, and Noriega, at the top of the P6 truss, pulled the slack tensioning cables through each of the reels. Tanner then turned each of the spring-loaded tension reels and let them unwind, allowing Noriega to guide the cables onto the reel grooves, thereby adding tension to the slack blanket. Once they repaired the tension problems with the solar array, Noriega and Tanner turned to their scheduled tasks of installing a centerline camera cable outside of Unity and an FPP at the top of P6 that would measure the electrical potential of plasma around ISS. The efficiency of the spacewalk allowed Tanner and Noriega time to complete tasks scheduled for spacewalks in the following month. The additional tasks included installing a sensor on the radiator, installing some small antennas, and photographically surveying the Station.[10]

11 December STS-97/Endeavour landing

10. Day 8 Highlights," National Aeronautics and Space Administration, accessed 11 January 2014, (*http://science.ksc.nasa.gov/shuttle/missions/sts-97/sts-97-day-08-highlights.html*); *STS-97, Mission Control Center: Status Report #15*, NASA Johnson Space Center, accessed 11 January 2014, (*http://spaceflight.nasa.gov/spacenews/reports/sts97/STS-97-15.html*).

2001

7 February STS-98/Atlantis launch

10 FEBRUARY

2001 EVA 1	**Duration:** 7:33
World EVA 192	**Spacecraft/mission:** STS-98
U.S. EVA 99	**Crew:** Kenneth Cockrell, Mark Polansky, Robert Curbeam, Thomas Jones, Marsha Ivins
Shuttle EVA 57	**Spacewalkers:** Thomas Jones, Robert Curbeam
	Purpose: First of three scheduled EVAs. Connect Destiny to ISS's power and cooling lines

Initially, the objectives were to install United States Laboratory Module Destiny and connect it to the ISS's electrical, data, and cooling lines. As Robert Curbeam attached a cooling line, a small amount of frozen ammonia crystals leaked into the ISS. To avoid contamination of Atlantis, Curbeam remained in the Sun for 30 minutes to vaporize any ammonia crystals. Meanwhile Thomas Jones brushed off the spacesuits and equipment of any visible ammonia crystals. The spacewalkers partially pressurized and then vented the Shuttle's airlock in case any ammonia remained in the air before final repressurization and re-entry into Atlantis. After the EVA, Kenneth Cockrell, Marsha Ivins, and Mark Polansky wore oxygen masks for 20 minutes to avoid any possible contamination from ammonia that may have remained in the airlock.[1]

12 FEBRUARY

2001 EVA 2	**Duration:** 6:50
World EVA 193	**Spacecraft/mission:** STS-98
U.S. EVA 100	**Crew:** Kenneth Cockrell, Mark Polansky, Robert Curbeam, Thomas Jones, Marsha Ivins
Shuttle EVA 58	**Spacewalkers:** Thomas Jones, Robert Curbeam
	Purpose: Second of three scheduled EVAs. Detach PMA-2 from Z1 truss; attach PMA-2 to Destiny

The Pressurized Mating Adapter 2 (PMA-2) was attached to the Z1 truss. Jones and Curbeam directed Marsha Ivins as she operated the RMS and removed the PMA-2 from the Z1 truss. The spacewalkers

1. "Mission Archives: STS-98," Jeanne Ryba, editor, accessed 20 December 2013, (*http://www.nasa.gov/mission_pages/shuttle/shuttlemissions/archives/sts-98.html*); "STS-98: a Destiny for the International Space Station," Kim Dismukes, editor, accessed 20 December 2013, (*http://spaceflight.nasa.gov/shuttle/archives/sts-98/index.html*).

continued to give visual cues to Ivins as she attached the Destiny module and the PMA-2. They connected computer and electrical cables between the port and the lab. The astronauts unveiled Destiny's large, high-quality window and installed its shutter. The window allowed the capture of high-quality videos and photographs. EVA handrails, sockets, and the base for the Space Station Remote Manipulator System (SSRMS) were also attached for future EVAs.[2]

14 FEBRUARY

2001 EVA 3	**Duration:** 5:25
World EVA 194	**Spacecraft/mission:** STS-98
U.S. EVA 101	**Crew:** Kenneth Cockrell, Mark Polansky, Robert Curbeam, Thomas Jones, Marsha Ivins
Shuttle EVA 59	**Spacewalkers:** Thomas Jones, Robert Curbeam
	Purpose: Third of three scheduled EVAs. Install spare S-band antenna; test incapacitated crew rescue

The third spacewalk required Jones and Curbeam to ensure successful cable connectivity between Destiny and PMA-2. They also attached an auxiliary S-band communications antenna and released the Station's cooling radiator. The astronauts examined the U.S. solar array. Near the end of the EVA, they practiced carrying an incapacitated crewmember to the airlock.[3]

20 February STS-98/Atlantis landing

8 March STS-102/Discovery/ISS Expedition 2 launch

2. "Mission Archives: STS-98," Jeanne Ryba, editor, accessed 20 December 2013, (*http://www.nasa.gov/mission_pages/shuttle/shuttlemissions/archives/sts-98.html*); "Remote Manipulator System," NASA Johnson Space Center, accessed 20 December 2013, (*http://prime.jsc.nasa.gov/ROV/rms.html*); "STS-98: a Destiny for the International Space Station," Kim Dismukes, editor, accessed 20 December 2013, (*http://spaceflight.nasa.gov/shuttle/archives/sts-98/index.html*).

3. "Mission Archives: STS-98," Jeanne Ryba, editor, accessed 20 December 2013, (*http://www.nasa.gov/mission_pages/shuttle/shuttlemissions/archives/sts-98.html*); "STS-98: a Destiny for the International Space Station," Kim Dismukes, editor, accessed 20 December 2013, (*http://spaceflight.nasa.gov/shuttle/archives/sts-98/index.html*).

11 MARCH

2001 EVA 4	**Duration:** 8:55
World EVA 195	**Spacecraft/mission:** STS-102, International Space Station Expeditions 1 and 2
U.S. EVA 102	**Crew:** James Wetherbee, James Kelly, Andrew Thomas, Paul Richards
Space Station EVA 106	**Expedition Crew 1:** William Shepherd, Sergei Krikalev (Russian Space Agency), Yuri Gidzenko (Russian Space Agency)
ISS EVA 1	**Expedition Crew 2:** James Voss, Susan Helms, Yury Usachev (Russian Space Agency)
	Spacewalkers: James Voss, Susan Helms
	Purpose: First of two scheduled EVAs. Disconnect PMA-3 from Node 1; berth of Leonardo to ISS

The Italian-built Leonardo was the first of three Multi-Purpose Logistics Modules installed on the ISS. They could transfer 10 tons of cargo between the Station and space shuttles. The 21 by 15-foot (6.40 by 4.57-meter) module carried electrical equipment, life support, experiments, and supplies. By disconnecting the PMA-3 from Unity, the ISS had space to berth Leonardo. James Voss and Susan Helms' record-breaking EVA was the longest in Shuttle history.[4]

13 MARCH

2001 EVA 5	**Duration:** 6:21
World EVA 196	**Spacecraft/mission:** STS-102, International Space Station Expeditions 1 and 2
U.S. EVA 103	**Crew:** James Wetherbee, James Kelly, Andrew Thomas, Paul Richards
Space Station EVA 107	**Expedition Crew 1:** William Shepherd, Sergei Krikalev (Russian Space Agency), Yuri Gidzenko (Russian Space Agency)
ISS EVA 2	**Expedition Crew 2:** James Voss, Susan Helms, Yury Usachev (Russian Space Agency)
	Spacewalkers: Andrew Thomas, Paul Richards
	Purpose: Second of two scheduled EVAs. Prepare the installation of a robotic arm

---

4. "International Space Station: Multi-Purpose Logistics Modules," John Ira Petty, editor, accessed 21 December 2013, (*http://www.nasa.gov/mission_pages/station/structure/elements/mplm.html*); "Mission Archives: STS-102," Jeanne Ryba, editor, accessed 21 December 2013, (*http://www.nasa.gov/mission_pages/shuttle/shuttlemissions/archives/sts-102.html*); "STS-102: a new crew of the International Space Station," Kim Dismukes, editor, accessed 21 December 2013, (*http://spaceflight.nasa.gov/shuttle/archives/sts-102/index.html*).

Paul Richards and Andrew Thomas completed deferred tasks from the previous spacewalk. First, they coupled power cables on the Lab Cradle Assembly that would berth Destiny's unassembled robotic arm. Next, the pair installed the External Stowage Platform that would eventually house both heating power to equipment storage and a Pump and Flow Control Subassembly to regulate ammonia coolant flow.[5]

21 March STS-102/Discovery/Expedition 1 landing

23 March Unmanned Space Station Mir reentry into the Pacific Ocean

19 April STS-100/Endeavour launch

22 APRIL

2001 EVA 6	**Duration:** 7:10
World EVA 197	**Spacecraft/mission:** STS-100
Canadian EVA 1/ U.S. EVA 104	**Crew:** Kent Rominger, Jeffrey Ashby, Scott Parazynski, John Phillips, Yuri Lonchakov (Russian Space Agency), Umberto Guidoni (European Space Agency), Chris Hadfield (Canadian Space Agency)
Shuttle EVA 60	
	Spacewalkers: Chris Hadfield, Scott Parazynski
	Purpose: First of two scheduled EVAs. Install Canadarm2 and its cables

Canadarm2 is a highly developed, 57.7-foot (17.58-meter) robotic arm secured to Destiny by Scott Parazynski and the first Canadian spacewalker, Chris Hadfield. The pair connected communication cables, giving Canadarm2 contact with the United States laboratory and ground controllers. The robotic arm can be remotely operated to transfer large payloads and spacewalkers.[6]

5. "Mission Archives: STS-102," Jeanne Ryba, editor, accessed 21 December 2013, (*http://www.nasa.gov/mission_pages/shuttle/shuttlemissions/archives/sts-102.html*); "STS-102: a new crew for the International Space Station," Kim Dismukes, editor, accessed 21 December 2013, (*http://spaceflight.nasa.gov/shuttle/archives/sts-102/index.html*); "STS-102 Day 5 Highlights," Kennedy Space Center, accessed 21 December 2013, (*http://science.ksc.nasa.gov/shuttle/missions/sts-102/sts-102-day-05-highlights.html*).

6. "Mission Archives: STS-100," Jeanne Ryba, editor, accessed 22 December 2013, (*http://www.nasa.gov/mission_pages/shuttle/shuttlemissions/archives/sts-100.html*); "Space Station Assembly: Canadarm2 and the Mobile Servicing System," Amiko Kauderer, editor, accessed 22 December 2013, (*http://www.nasa.gov/mission_pages/station/structure/elements/mss.html*); "STS-100 Day-4 Highlights," Kennedy Space Center, accessed 22 December 2013, (*http://science.ksc.nasa.gov/shuttle/missions/sts-100/sts-100-day-04-highlights.html*).

24 APRIL

2001 EVA 7	**Duration:** 7:40
World EVA 198	**Spacecraft/mission:** STS-100
Canadian EVA 2/ U.S. EVA 105	**Crew:** Kent Rominger, Jeffrey Ashby, Scott Parazynski, John Phillips, Yuri Lonchakov (Russian Space Agency), Umberto Guidoni (European Space Agency), Chris Hadfield (Canadian Space Agency)
Shuttle EVA 61	**Spacewalkers:** Chris Hadfield, Scott Parazynski
	Purpose: Second of two scheduled EVAs. Disconnect cables between lab and SSRMS; remove and stow ECOMM antenna

The Canadarm2, also known as the SSRMS, was connected to Destiny lab via the Power and Data Grapple Fixture. With the help of Susan Helms in the lab, Hadfield and Parazynski reconnected the arm's cables until a reliable power path to the arm was created. The SSRMS was designed to "walk" on the ISS by grappling external fixtures. The spacewalkers also detached an ECOMM antenna and equipped a spare Direct Current Switching Unit from Endeavour to Destiny's storage rack.[7]

28 April Soyuz-TM 32 landing

1 May STS-100/Endeavour landing

6 May Soyuz-TM 31 landing

8 JUNE

2001 EVA 8	**Duration:** 0:19
World EVA 199	**Spacecraft/mission:** International Space Station Expedition 2
Russian EVA 99/ U.S. EVA 106	**Crew:** James Voss, Susan Helms, Yury Usachev (Russian Space Agency)
Space Station EVA 108	**Spacewalkers:** Yury Usachev, James Voss
ISS EVA 3	**Purpose:** Relocate docking cone; Test Orlan suits

The first Station residents to execute an EVA from the ISS were Yury Usachev and James Voss. Without a docked Shuttle to provide assistance, the spacewalkers only had support from Susan Helms in the Zarya module. The short, 19-minute internal EVA was performed inside the Service Module transfer compartment and required attaching the Zvezda hatch cover to the top of the module and replacing its docking cone.[8]

---

7. "Mission Archives: STS-100," Jeanne Ryba, editor, accessed 22 December 2013, (*http://www.nasa.gov/mission_pages/shuttle/shuttlemissions/archives/sts-100.html*); "Space Station Assembly: Canadarm2 and the Mobile Servicing System," Amiko Kauderer, editor, accessed 10 December 2015, (*http://www.nasa.gov/mission_pages/station/structure/elements/mss.html*).

8. "International Space Station Status Report #01-18,"Johnson Space Center, accessed 23 December 2013, (*http://www.nasa.gov/centers/johnson/news/station/2001/iss01-18.html*).

12 July STS-104/Atlantis launch

14 JULY

2001 EVA 9	**Duration:** 5:59
World EVA 200	**Spacecraft/mission:** STS-104
U.S. EVA 107	**Crew:** Steven Lindsey, Charles Hobaugh, Michael Gernhardt, James Reilly, Janet Kavandi
Shuttle EVA 62	**Spacewalkers:** Michael Gernhardt, James Reilly
	Purpose: First of three scheduled EVAs. Install Quest Airlock to Unity

Previously, the ISS Expedition 2 crew removed the Zvezda hatch to prepare for the berthing of the new Quest airlock. The STS-104 crew delivered the airlock, which supports both EMUs and Russian Orlan spacesuits during spacewalk exits and entries. As Michael Gernhardt and James Reilly removed the airlock's insulating covers, Helms lifted the airlock from Atlantis' payload bay with Canadarm2. The spacewalkers guided Helms as she planted the airlock onto Unity's berthing port. Gernhardt then coupled the ISS's heating cables to the airlock.[9]

17 JULY

2001 EVA 10	**Duration:** 6:26
World EVA 201	**Spacecraft/mission:** STS-104
U.S. EVA 108	**Crew:** Steven Lindsey, Charles Hobaugh, Michael Gernhardt, James Reilly, Janet Kavandi
Shuttle EVA 63	**Spacewalkers:** Michael Gernhardt, James Reilly
	Purpose: Second of three scheduled EVAs. Install three of four Quest Airlock tanks

Gernhardt and Reilly removed three of the Joint Airlock's high-pressure gas tanks from Atlantis's Spacelab pallet. With the Station's Canadarm2 and the Shuttle's Canadarm, Gernhardt and Reilly outfitted a nitrogen tank and two oxygen tanks. Once installed, the airlock prevented major loss of air during spacewalkers' airlock egress and ingress.[10]

---

9. "International Space Station: Quest Airlock," Jerry Wright, editor, accessed 2 January 2014, (*http://www.nasa.gov/mission_pages/station/structure/elements/quest.html*); "Mission Archive: STS-104," Jeanne Ryba, editor, accessed 2 January 2014, (*http://www.nasa.gov/mission_pages/shuttle/shuttlemissions/archives/sts-104.html*); "STS-104: giving the space station a doorway to space," Kim Dismukes, editor, accessed 2 January 2014, (*http://spaceflight.nasa.gov/shuttle/archives/sts-104/index.html*).

10. "Mission Archive: STS-104," Jeanne Ryba, editor, accessed 2 January 2014, (*http://www.nasa.gov/mission_pages/shuttle/shuttlemissions/archives/sts-104.html*); "STS-104," Kennedy Space Center, accessed 3 January 2014, (*http://science.ksc.nasa.gov/shuttle/missions/sts-104/mission-sts-104.html*)

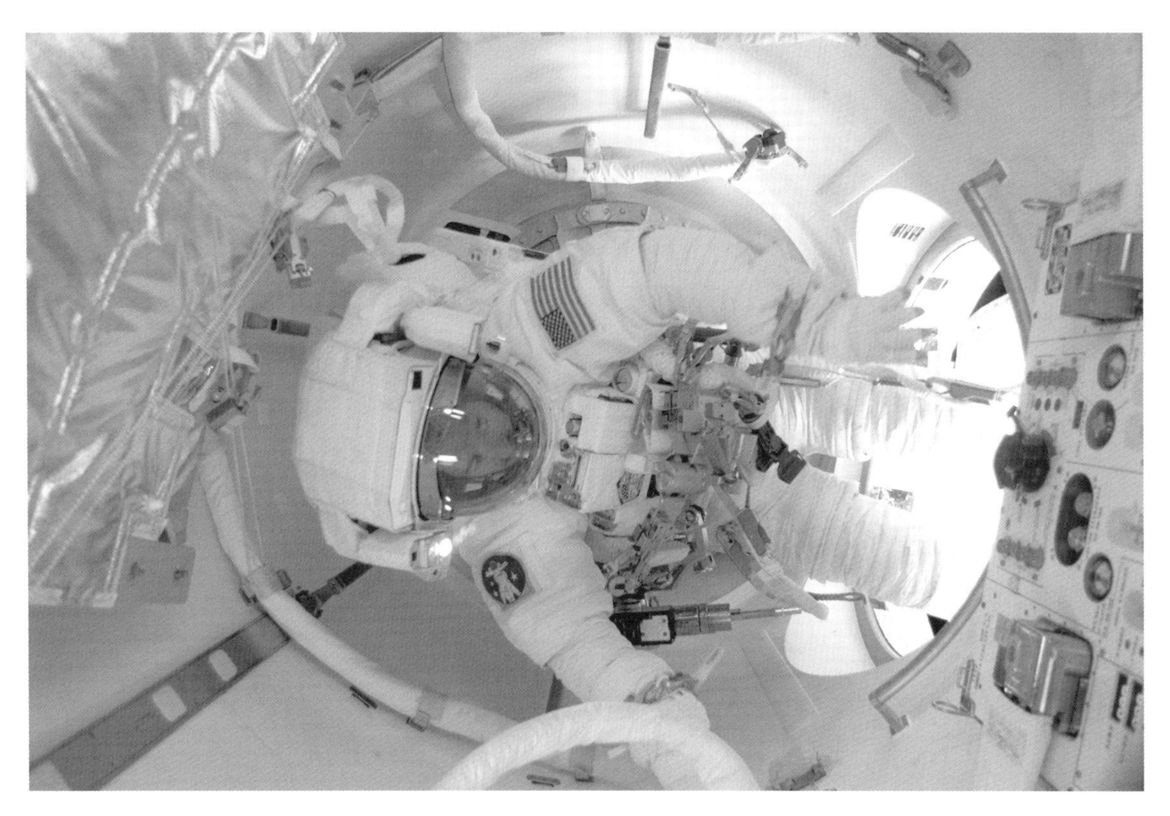

FIGURE 4. **First use of the Quest airlock.** Astronaut James F. Reilly, STS-104 mission specialist, participates in space history as he joins fellow astronaut and mission specialist Michael L. Gernhardt (out of frame) in utilizing the new Quest airlock for the first ever spacewalk to egress from the ISS. (NASA STS104-E-5237)

20 JULY

2001 EVA 11	**Duration:** 4:01
World EVA 202	**Spacecraft/mission:** STS-104
U.S. EVA 109	**Crew:** Steven Lindsey, Charles Hobaugh, Michael Gernhardt, James Reilly, Janet Kavandi
Shuttle EVA 64	**Spacewalkers:** Michael Gernhardt, James Reilly
	Purpose: Third of three scheduled EVAs. Complete nitrogen tank installation; mount handrails and stow EVA tools

The pair became the first spacewalkers to enter and exit from the Quest airlock. Gernhardt and Reilly's assembly of the last nitrogen tank completed the installation of the High Pressure Gas Tank. The tanks pressurize the airlock and supply oxygen to the spacesuits. Gernhardt and Reilly completed Quest's assembly by mounting handrails and other EVA tools to its exterior.[11]

---

11. "International Space Station: Quest Airlock," Jerry Wright, editor, accessed 2 January 2014, (*http://www.nasa.gov/mission_pages/station/structure/elements/quest.html*); "Mission Archive: STS-104," Jeanne Ryba, editor, accessed 2 January 2014, (*http://www.nasa.gov/mission_pages/shuttle/shuttlemissions/archives/sts-104.html*); "STS-104: giving the space station a doorway to space," Kim Dismukes, editor, accessed 2 January 2014, (*http://spaceflight.nasa.gov/shuttle/archives/sts-104/index.html*).

24 July STS-104/Atlantis landing

10 August STS-105/Discovery/ISS Expedition 3 launch

16 AUGUST

2001 EVA 12	**Duration:** 6:16
World EVA 203	**Spacecraft/mission:** STS-105
U.S. EVA 110	**Crew:** Scott Horowitz, Frederick Sturckow, Daniel Barry, Patrick Forrester Frank Culbertson (began tour on ISS), and Mikhail Tyurin and Vladimir Dezhurov (Russian Space Agency, began tour on ISS); James Voss, Yury Usachev, and Susan Helms (returned to Earth)
Shuttle EVA 65	**Spacewalkers:** Daniel Barry, Patrick Forrester
	Purpose: First of two scheduled EVAs. Install Early Ammonia Servicer (EAS) on P6 truss; attach two Materials International Space Station Equipment (MISSE) to airlock exterior

This was the first of Daniel Barry and Patrick Forrester's two spacewalks devoted to the construction of the ISS. The pair first installed an EAS on the Station. The ammonia within the EAS is to be used as a coolant for the Station's radiator. Barry and Forrester also initiated the MISSE, where they attached two containers of material samples to the Station's hull. The experiment consisted of 750 samples of different materials left exposed in space for roughly 18 months before being recollected and returned to Earth for study.[12]

18 AUGUST

2001 EVA 13	**Duration:** 5:29
World EVA 204	**Spacecraft/mission:** STS-105
U.S. EVA 111	**Crew:** Scott Horowitz, Frederick Sturckow, Daniel Barry, Patrick Forrester
Shuttle EVA 66	**Spacewalkers:** Daniel Barry, Patrick Forrester
	Purpose: Second of two scheduled EVAs. Install handrails on Destiny; attach heater cables on lab for later use with the Starboard Zero (S0) truss

Barry and Forrester fitted handrails and strung two 45-foot (13.71-meter) heater cables down both sides of the Destiny laboratory. The cables were hung as a precautionary measure to supply the S0 truss with power in the event that during its delivery the following year it could not be installed promptly and properly.[13]

---

12. "STS-105 Day 7 Highlights," National Aeronautics and Space Administration, accessed 8 January 2014, (*http://science.ksc.nasa.gov/shuttle/missions/sts-105/sts-105-day-07-highlights.html*); *STS-105, Mission Control Center: Status Report #13*, NASA Johnson Space Center, accessed 8 January 2014, (*http://spaceflight.nasa.gov/spacenews/reports/sts105/STS-105-13.html*).

13. "STS-105 Day 9 Highlights," National Aeronautics and Space Administration, accessed 8 January 2014, (*http://science.ksc.nasa.gov/shuttle/missions/sts-105/sts-105-day-09-highlights.html*); *STS-105, Mission Control Center: Status Report #17*, NASA Johnson Space Center, accessed 8 January 2014, (*http://spaceflight.nasa.gov/spacenews/reports/sts105/STS-105-17.html*).

22 August STS-105/Discovery/ISS Expedition 2 landing

8 OCTOBER

2001 EVA 14	**Duration:** 4:58
World EVA 205	**Spacecraft/mission:** International Space Station Expedition 3
Russian EVA 100	**Crew:** Vladimir Dezhurov, Mikhail Tyurin, Frank Culbertson
Space Station EVA 109	**Spacewalkers:** Vladimir Dezhurov, Mikhail Tyurin
ISS EVA 4	**Purpose:** Connect cables between Zvezda and Pirs; mount vital spacewalk equipment; test Strela cargo crane

Vladimir Dezhurov and Mikhail Tyurin were the first to exit from the Russian airlock without a docked Shuttle. The cosmonauts connected communication and data cables between the Zvezda Service Module and the Pirs Docking Compartment, which is also known as DC-1. Pirs is both a docking port for cargo vehicles and an airlock for spacewalkers in Orlan spacesuits. The duo also mounted an egress/ingress ladder, four handrails, a Kurs antenna, a Strela cargo crane, and a docking target to Pirs' exterior. They did not, however, have enough time to finalize testing of the Strela cargo crane.[14]

15 OCTOBER

2001 EVA 15	**Duration:** 5:52
World EVA 206	**Spacecraft/mission:** International Space Station Expedition 3
Russian EVA 101	**Crew:** Vladimir Dezhurov, Mikhail Tyurin, Frank Culbertson
Space Station EVA 110	**Spacewalkers:** Vladimir Dezhurov, Mikhail Tyurin
ISS EVA 5	**Purpose:** Install Kromka to Zvezda; replace Russian flag placard with Kodak sign

The cosmonauts installed three sets of experiments inside the Pirs Docking Compartment. The supplies were later used to study the space environment surrounding the Station. Dezhurov and Tyurin then installed Kromka, a device measuring contamination from steering jets, behind Zvezda. Kromka's collected data were later applied to improving future thrusters. The spacewalkers assembled a small truss and three suitcase-sized experiments contracted by the National Space Development Agency of Japan (NASDA). The research required the Micro-Particle Capturer to collect micrometeoroids and human-made particles with aerogel and foam substances. In addition, the Space Environment Exposure Device would allow the study of paint, insulation, and lubricants in space. As they returned to the airlock, Dezhurov and Tyurin replaced the Russian flag placard with a Kodak sign as part of a

14. "International Space Station: Pirs," Jerry Wright, editor, accessed 8 January 2014, (*http://www.nasa.gov/mission_pages/station/structure/elements/pirs.html*); "International Space Station Status Report #01-34," NASA Johnson Space Center, accessed 8 January 2014, (*http://www.nasa.gov/centers/johnson/news/station/2001/iss01-34.html*).

FIGURE 5. **First Soyuz Docking on Pirs.** A Soyuz spacecraft approaches the ISS carrying the Soyuz taxi crew, Commander Victor Afanasyev, Flight Engineer Konstantin Kozeev, and French Flight Engineer Claudie Haignere for an 8-day stay on the station. (NASA ISS003-E-6849)

commercial agreement. As the duo worked, Frank Culbertson maneuvered the Canadarm2 for ground control visuals in Houston and Moscow.[15]

21 October Soyuz-TM 33 launch

31 October Soyuz-TM 32 landing

12 NOVEMBER

2001 EVA 16	**Duration:** 5:04
World EVA 207	**Spacecraft/mission:** International Space Station Expedition 3
Russian EVA 102/ U.S. EVA 112	**Crew:** Vladimir Dezhurov, Mikhail Tyurin, Frank Culbertson
	Spacewalkers: Vladimir Dezhurov, Frank Culbertson
Space Station EVA 111	**Purpose:** Connect Kurs cables between Zvezda and Pirs; test Strela crane; photograph and examine Zvezda solar array panels
ISS EVA 6	

15. "International Space Station Status Report #01-36," John Ira Petty, editor, accessed 8 January 2014, (*http://www.nasa.gov/centers/johnson/news/station/2001/iss01-36.html*); "Space Environment Data Acquisition Equipment," Victor M. Escobedo Jr., editor, accessed 8 January 2014, (*http://www.nasa.gov/mission_pages/station/research/experiments/671.html*).

Frank Culbertson and Dezhurov successfully coupled four high frequency and three low frequency communication cables between Zvezda and Pirs. The Kurs automated rendezvous system was set to lead Russian vehicles to the Pirs docking port. The pair also inspected Zvezda's solar array panels that failed to deploy during its launch on July 12, 2000. The panel remained safe to the crew. They completed the EVA after testing the Russian Strela cargo crane.[16]

3 DECEMBER

2001 EVA 17	**Duration:** 2:46
World EVA 208	**Spacecraft/mission:** International Space Station Expedition 3
Russian EVA 103	**Crew:** Vladimir Dezhurov, Mikhail Tyurin, Frank Culbertson
Space Station EVA 112	**Spacewalkers:** Vladimir Dezhurov, Mikhail Tyurin
ISS EVA 7	**Purpose:** Remove debris on docking port; photograph and remove Progress 5 debris

Progress's initial attempt to dock was prevented by debris around the Zvezda docking port. Dezhurov and Tyurin removed the blockage allowing a successful dock. A week earlier, Endeavour's launch was rescheduled due to debris from a previous Progress supply vehicle.[17]

5 December STS-108/Endeavour/ISS Expedition 4 launch

10 DECEMBER

2001 EVA 18	**Duration:** 4:12
World EVA 209	**Spacecraft/mission:** STS-108
U.S. EVA 113	**Crew:** Dominic Gorie, Mark Kelly, Linda Godwin, Daniel Tani, Yuri Onufrienko (Russian Space Agency), Mikhail Tyurin (Russian Space Agency), Vladimir Dezhurov (Russian Space Agency)
Shuttle EVA 67	**Spacewalkers:** Linda Godwin, Daniel Tani
	Purpose: Install insulation on ISS's Beta Gimbal Assemblies and retrieval of antenna cover

Linda Godwin and Daniel Tani climbed atop the P6 truss from Endeavour's cargo bay to place insulator blankets over two Beta Gimbal Assemblies, which rotate the ISS's solar arrays so that they can track the Sun's rays. Once this was completed, they attempted to secure one of the four legs that hold the starboard array to the Station, but were unable to close its latch. On their way down from P6,

16. "International Space Station Status Report #01-43," NASA Johnson Space Center, accessed 8 January 2014, (*http://www.nasa.gov/centers/johnson/news/station/2001/iss01-43.html*).

17. "International Space Station Status Report #01-49," NASA Johnson Space Center, accessed 8 January 2014, (*http://www.nasa.gov/centers/johnson/news/station/2001/iss01-49.html*).

Godwin and Tani also retrieved a cover originally from a Station antenna in a stowage bin. Before reentering Endeavour, the two positioned two switches on the Station's exterior to install during the STS-110 mission.[18]

17 December STS-108/Endeavour/ISS Expedition 3 landing

18. "Mission Archives: STS-108," Jeanne Ryba, editor, accessed 11 January 2014, (*http://www.nasa.gov/mission_pages/shuttle/shuttlemissions/archives/sts-108.html*); "STS-108 Day 6 Highlights," National Aeronautics and Space Administration, accessed 11 January 2014, (*http://science.ksc.nasa.gov/shuttle/missions/sts-108/sts-108-day-06-highlights.html*).

2002

14 JANUARY

2002 EVA 1	**Duration:** 6:03
World EVA 210	**Spacecraft/mission:** International Space Station Expedition 4
Russian EVA 104/ U.S. EVA 114	**Crew:** Daniel Bursch, Carl Walz, Yuri Onufrienko (Russia Space Agency)
	Spacewalkers: Yuri Onufrienko, Carl Walz
Space Station EVA 113	**Purpose:** Install Strela cargo crane to Pirs; install amateur radio antenna
ISS EVA 8	

Yuri Onufrienko and Carl Walz completed the installation of a second Strela cargo crane, known as Strela 2. Using the functioning Strela 1 cargo crane, the spacewalkers first removed Strela 2 from its 30-month storage between the Unity and Zarya tunnel, known as PMA-1. They then attached Strela 2 to Pirs and finalized its installation. The two cargo cranes would later maneuver spacewalkers and their equipment during EVAs. Onufrienko and Walz also mounted the first of four amateur communications antennas on Zvezda, with the help of Flight Engineer Daniel Bursch inside.[1]

25 JANUARY

2002 EVA 2	**Duration:** 5:59
World EVA 211	**Spacecraft/mission:** International Space Station Expedition 4
Russian EVA 105/ U.S. EVA 115	**Crew:** Daniel Bursch, Carl Walz, Yuri Onufrienko (Russia Space Agency)
	Spacewalkers: Yuri Onufrienko, Dan Bursch
Space Station EVA 114	**Purpose:** Install six thruster deflector shields; install amateur radio antenna
ISS EVA 9	

The Canadarm2 television cameras allowed Walz to observe and direct the spacewalk from inside the Station. Meanwhile in Russian Orlan spacesuits, Bursch and Onufrienko first focused on the Zvezda attitude control thrusters. The duo installed six plume deflectors on Zvezda, allowing regulation of the sediments released into space by the thrusters. Then, they began work on five experiments attached to Zvezda. The spacewalkers first retrieved the Kromka experiment, which recorded the volume and types of deposits released by the thrusters. The experiment returned to Earth in May 2002 for further research. Bursch and Onufrienko also mounted a new Kromka experiment identical to the one they had just collected. Then, they mounted the Platan-M, which photographed low-energy heavy particles

1. "International Space Station Status Report #02-03," John Ira Petty, editor, accessed 15 February 2014, (*http://www.nasa.gov/centers/johnson/news/station/2002/iss02-03.html*); *Praxis Manned Spaceflight Log: 1961–2006*, Tim Furniss and David J. Shayler with Michael D. Shayler, Springer, 2007, p. 702; *Russian Spacesuits,* Isaak P. Abramov and Å. Ingemar Skoog, Springer, 2003, p. 316.

from the Sun. Next, Bursch and Onufrienko installed three related Russian experiments that recorded the effects space had on numerous materials. These experiments are known as the Replaceable Cassette Container (SKK) experiments. Furthermore, the duo installed fairleads on Zvezda handrails to prevent collisions between tethered equipment. The spacewalkers also continued a task from the previous spacewalk and installed the second amateur radio antenna and its cables on Zvezda.[2]

20 FEBRUARY

2002 EVA 3	**Duration:** 5:47
World EVA 212	**Spacecraft/mission:** International Space Station Expedition 4
U.S. EVA 116	**Crew:** Daniel Bursch, Carl Walz, Yuri Onufrienko (Russia Space Agency)
Space Station EVA 115	**Spacewalkers:** Carl Walz, Dan Bursch
ISS EVA 10	**Purpose:** First use of the Quest Airlock without Shuttle docked; prepare installation of S0 truss; relocate tools and equipment for future spacewalks

On the 40th anniversary of John Glenn's first flight, astronauts Bursch and Walz performed the first solely U.S. spacewalk from the ISS using the Quest airlock. Furthermore, they were the first to exit the Station's Quest airlock without a docked Shuttle in support. Astronaut Joe Tanner assisted from Mission Control as the excursion's Capsule Communicator (CAPCOM). Bursch and Walz's objective was to prepare the installation of the S0 truss, which was later completed in April 2002. The spacewalkers arranged the tools, cables, and adaptors necessary for the four future EVAs to assemble the truss.[3]

1 March STS-109/Columbia launch

4 MARCH

2002 EVA 4	**Duration:** 7:01
World EVA 213	**Spacecraft/mission:** STS-109
U.S. EVA 117	**Crew:** Scott Altman, Duane Carey, John Grunsfeld, Nancy Currie, James Newman, Richard Linnehan, Michael Massimino
Shuttle EVA 68	**Spacewalkers:** John Grunsfeld, Richard Linnehan
	Purpose: Hubble Servicing Mission 3B, the first of five scheduled EVAs

2. "International Space Station Status Report #02-05," John Ira Petty, editor, accessed 15 February 2014, (*http://www.nasa.gov/centers/johnson/news/station/2002/iss02-05.html*); *Praxis Manned Spaceflight Log: 1961–2006*, Tim Furniss and David J. Shayler with Michael D. Shayler, Springer, 2007, p. 702; *Russian Spacesuits,* Isaak P. Abramov and Å. Ingemar Skoog, Springer, 2003, p. 316.

3. "Integrated Truss Structure," John Ira Petty, editor, accessed 15 February 2014, (*http://www.nasa.gov/centers/johnson/news/station/2002/iss02-10.html*); "International Space Station Status Report #02-10," John Ira Petty, editor, accessed 15 February 2014, (*http://www.nasa.gov/centers/johnson/news/station/2002/iss02-10.html*); *Praxis Manned Spaceflight Log: 1961–2006*, Tim Furniss and David J. Shayler with Michael D. Shayler, Springer, 2007, p. 702; *Russian Spacesuits,* Isaak P. Abramov and Å. Ingemar Skoog, Springer, 2003, p. 316.

During their 11-day mission, the STS-109 crew focused on servicing the Hubble Space Telescope. The crew dedicated five spacewalks to rejuvenating and updating HST. Before the spacewalks began, they grasped the telescope with Columbia's robotic arm and berthed it inside the payload bay. John Grunsfeld and Richard Linnehan then performed the first of five EVAs. The spacewalkers tethered themselves to the robotic arm, allowing Nancy Currie to maneuver them around the telescope. Grunsfeld and Linnehan replaced the old starboard solar array with a smaller, more power generating third-generation array. The old solar array was returned to Earth, and its 9-year performance on HST was studied.[4]

5 MARCH

2002 EVA 5	**Duration:** 7:16
World EVA 214	**Spacecraft/mission:** STS-109
U.S. EVA 118	**Crew:** Scott Altman, Duane Carey, John Grunsfeld, Nancy Currie, James Newman, Richard Linnehan, Michael Massimino
Shuttle EVA 69	**Spacewalkers:** James Newman, Michael Massimino
	Purpose: Hubble Servicing Mission 3B, the second of five scheduled EVAs

During the second of five planned EVAs, James Newman and Michael Massimino first replaced the other old HST solar array with the new smaller array. They then installed a Reaction Wheel Assembly, which allows remote reorientation of the telescope. Similar to the previous day's EVA, the spacewalkers used Columbia's robotic arm to maneuver around the telescope. Time permitted Newman and Massimino to accomplish additional tasks. They positioned foot restraints in preparation for Grunsfeld and Linnehan's spacewalk the following day, and installed thermal blankets on Bay 6 and doorstop extensions on Bay 5. The bays are HST compartments that house various electronic utilities. Furthermore, they tested the telescope's two aft shroud doors and replaced two bottom bolts.[5]

6 MARCH

2002 EVA 6	**Duration:** 6:48
World EVA 215	**Spacecraft/mission:** STS-109
U.S. EVA 119	**Crew:** Scott Altman, Duane Carey, John Grunsfeld, Nancy Currie, James Newman, Richard Linnehan, Michael Massimino
Shuttle EVA 70	**Spacewalkers:** John Grunsfeld, Richard Linnehan
	Purpose: Hubble Servicing Mission 3B, the third of five scheduled EVAs

---

4. "Mission Archives: STS-109," Jeanne Ryba, editor, accessed 15 February 2014, (*http://www.nasa.gov/mission_pages/shuttle/shuttlemissions/archives/sts-109.html*); *Praxis Manned Spaceflight Log: 1961–2006*, Tim Furniss and David J. Shayler with Michael D. Shayler, Springer, 2007, p. 706.

5. Ibid.

The EVA was delayed due to a water leak in Grunsfeld's spacesuit. He replaced the upper portion of the suit, known as the Hard Upper Torso (HUT), and was then able to perform the spacewalk with Linnehan. Because the new solar arrays generated 20 percent more power, the telescope required a new Power Control Unit (PCU) that could handle the increase. The original PCU was launched with the telescope 12 years earlier, in 1990. The Space Telescope Operations Control Center in Greenbelt, Maryland, powered down HST for the first time since its launch, allowing the spacewalkers to replace the PCU. Linnehan first disconnected 30 of 36 PCU connectors. Grunsfeld unhooked the remaining six and brought the old PCU to the payload bay. He then installed the telescope's new PCU and mated its cables. An hour later, the mission was proven successful with the telescope's power-up.[6]

7 MARCH

2002 EVA 7
World EVA 216
U.S. EVA 120
Shuttle EVA 71

Duration: 7:30
Spacecraft/mission: STS-109
Crew: Scott Altman, Duane Carey, John Grunsfeld, Nancy Currie, James Newman, Richard Linnehan, Michael Massimino
Spacewalkers: James Newman, Michael Massimino
Purpose: Hubble Servicing Mission 3B, the fourth of five scheduled EVAs

Newman and Massimino replaced Hubble's Faint Object Camera with the Advanced Camera for Surveys (ACS), marking the mission's first science instrument upgrade and the removal of the only remaining original instrument. The ACS was 10 times more efficient than the previous Hubble camera. The new camera was used to photograph many of the famous images using the telescope, most notably the Hubble Ultra Deep Field. After replacing the Faint Object Camera, Massimino installed the Electronic Support Module (ESM), the first element of an experimental cooling system. The remainder of the cooling system was installed the next day.[7]

8 MARCH

2002 EVA 8
World EVA 217
U.S. EVA 121
Shuttle EVA 72

Duration: 7:20
Spacecraft/mission: STS-109
Crew: Scott Altman, Duane Carey, John Grunsfeld, Nancy Currie, James Newman, Richard Linnehan, Michael Massimino
Spacewalkers: John Grunsfeld, Richard Linnehan
Purpose: Hubble Servicing Mission 3B, the fifth of five scheduled EVAs

---

6. "Mission Archives: STS-109," Jeanne Ryba, editor, accessed 20 February 2014, (*http://www.nasa.gov/mission_pages/shuttle/shuttlemissions/archives/sts-109.html*); *Praxis Manned Spaceflight Log: 1961–2006*, Tim Furniss and David J. Shayler with Michael D. Shayler, Springer, 2007, p. 706.

7. "Mission Archives: STS-109," Jeanne Ryba, editor, accessed 20 February 2014, (*http://www.nasa.gov/mission_pages/shuttle/shuttlemissions/archives/sts-109.html*); "Mission to Hubble: Hubble Space Telescope Servicing Mission 4," Lori Tyahla, editor, accessed 20 February 2014, (*http://www.nasa.gov/mission_pages/hubble/servicing/SM4/main/ACS_R_FS_HTML.html*); *Praxis Manned Spaceflight Log: 1961–2006*, Tim Furniss and David J. Shayler with Michael D. Shayler, Springer, 2007, p. 706.

During the crew's final spacewalk, Grunsfeld and Linnehan installed the Near Infrared Camera/Multi-Object Spectrometer (NICMOS) Cryocooler in the aft shroud. NICMOS was installed in February 1997 and designed to investigate dim, unknown, and unseen space. However, its solid nitrogen coolant was consumed more quickly than anticipated due to a heat leak. NICMOS reached cryogen exhaustion and remained dormant since January 3, 1999. During their spacewalk, Grunsfeld and Linnehan replaced the solid nitrogen coolant with the Cryocooler, allowing NICMOS to operate again. Furthermore, they installed the Cooling System Radiator to Hubble's exterior. Linnehan fed the radiator cables through the bottom of the telescope and Grunsfeld connected them to NICMOS. The next day, 9 March 2002, Columbia's robotic arm released HST. With 35 hours and 55 minutes required for five spacewalks, the STS-109 crew spent the most time servicing HST, more than any other Shuttle crew.[8]

12 March STS-109/Columbia landing

8 April STS-110/Atlantis launch

11 APRIL

2002 EVA 9	**Duration:** 7:48
World EVA 218	**Spacecraft/mission:** STS-110
U.S. EVA 122	**Crew:** Michael Bloomfield, Stephen Frick, Jerry Ross, Steven Smith, Ellen Ochoa, Lee Morin, Rex Walheim
Shuttle EVA 73	**Spacewalkers:** Steven Smith, Rex Walheim
	Purpose: First of four scheduled EVAs. Connect cables between Destiny and S0 truss

The STS-110 crew dedicated four spacewalks to mate the 13.5-ton S0 truss with the United States Destiny Laboratory. The S0 truss later served as the platform and junction for 10 other trusses and additional solar arrays on the ISS. Furthermore, the truss housed computers, navigation devices, and coolant and power sources for future modules. During this EVA, Steven Smith and Rex Walheim completed the initial steps to mount the truss and secure power cables to Destiny. They first bolted two of the four struts on Destiny and prepared avionics equipment and cables for the following spacewalks. The pair then connected an umbilical system between the truss and the Mobile Transporter. The latter was a significant asset that allowed the Canadarm2 to move around the Station. The Canadarm2 initially was stationary with a 50-foot (15.24-meter) reach. The Mobile Transporter runs on a railway along the S0 truss and served as a base for the Station's robotic arm. The Mobile Transporter

---

8. "Hubble Space Telescope NCS," Space Telescope Science Institute, accessed 20 February 2014, (*http://www.stsci.edu/hst/nicmos/documents/status_reports/nicmos_cryo.html*); "Mission Archives: STS-109," Jeanne Ryba, editor, accessed 20 February 2014, (*http://www.nasa.gov/mission_pages/shuttle/shuttlemissions/archives/sts-109.html*); "NICMOS Cryocooler-Reactivating a Hubble Instrument," NASA Goddard Space Flight Center, accessed 20 February 2014, (*http://ipp.nasa.gov/innovation/innovation104/6-smallbiz1.html*); *Praxis Manned Spaceflight Log: 1961–2006*, Tim Furniss and David J. Shayler with Michael D. Shayler, Springer, 2007, p. 706.

later assisted in installing the other trusses. Walheim was the first spacewalker to ride the mobile Canadarm2 and move equipment around the ISS.[9]

13 APRIL

2002 EVA 10	**Duration:** 7:30
World EVA 219	**Spacecraft/mission:** STS-110
U.S. EVA 123	**Crew:** Michael Bloomfield, Stephen Frick, Jerry Ross, Steven Smith, Ellen Ochoa, Lee Morin, Rex Walheim
Shuttle EVA 74	**Spacewalkers:** Jerry Ross, Lee Morin
	Purpose: Second of four scheduled EVAs. Mount struts on Destiny; remove S0 truss support clamps and panels

Jerry Ross and Lee Morin attached the final two struts on Destiny to mount the S0 truss. While Morin rode the Canadarm2 and Ross was tethered to the Station, the pair detached the truss's clamps and panels used for support during launch. They were unable to unscrew a bolt and left the task for a future spacewalk. The spacewalkers then installed a backup device for the Mobile Transporter.[10]

14 APRIL

2002 EVA 11	**Duration:** 6:27
World EVA 220	**Spacecraft/mission:** STS-110
U.S. EVA 124	**Crew:** Michael Bloomfield, Stephen Frick, Jerry Ross, Steven Smith, Ellen Ochoa, Lee Morin, Rex Walheim
Shuttle EVA 75	**Spacewalkers:** Steven Smith, Rex Walheim
	Purpose: Third of four scheduled EVAs. Release S0 truss from adapter latch; reconfigure cables between Destiny and Mobile Transporter; mount Airlock Spur to Quest Airlock

Smith and Walheim released the S0 truss from the adapter latch, a claw that restrained the truss to Destiny. The pair then reconfigured electrical connections between Destiny and the Mobile Transporter to power Canadarm2. While Walheim free-floated and Smith worked from the robotic arm, the pair

9. "Mission Archives: STS-110," Jeanne Ryba, editor, accessed 22 February 2014, (*http://www.nasa.gov/mission_pages/shuttle/shuttlemissions/archives/sts-110.html*); *Praxis Manned Spaceflight Log: 1961–2006*, Tim Furniss and David J. Shayler, Springer, 2007, p. 708; "The Slowest and Fastest Train in the Universe," Jeanne Ryba, editor, accessed 22 February 2014, (*http://www.nasa.gov/missions/shuttle/f_slowtrain.html*).

10. "Mission Archives: STS-110," Jeanne Ryba, editor, accessed 22 February 2014, (*http://www.nasa.gov/mission_pages/shuttle/shuttlemissions/archives/sts-110.html*); *Praxis Manned Spaceflight Log: 1961–2006*, Tim Furniss and David J. Shayler, Springer, 2007, p. 709.

unbolted the clamps that locked the Mobile Transporter to the truss. They delayed their final task, to attach the 14-foot (4.26-meter) Airlock Spur to the Quest Airlock, to the final EVA.[11]

16 APRIL

2002 EVA 12	**Duration:** 6:37
World EVA 221	**Spacecraft/mission:** STS-110
U.S. EVA 125	**Crew:** Michael Bloomfield, Stephen Frick, Jerry Ross, Steven Smith, Ellen Ochoa, Lee Morin, Rex Walheim
Shuttle EVA 76	**Spacewalkers:** Jerry Ross, Lee Morin
	Purpose: Fourth of four scheduled EVAs. Install S0-airlock handrail spur; install handrails and two Crew Equipment Translation Aid (CETA) lights on S0 truss; deploy Mobile Transporter energy absorbers.

During the mission's last spacewalk, Ross used the Canadarm2 and Morin free-floated while tethered to the ISS. They installed the Airlock Spur and prepared the S0 truss for future constructions. The Airlock Spur was moved from the S0 truss to the Quest Airlock. It served as a pathway for future spacewalkers assembling trusses. The pair then installed floodlights on Unity and Destiny, providing illumination during spacewalks. The spacewalkers also mounted a work platform for future constructions, installed electrical converters and circuit breakers, and connected shock absorbers to the Mobile Transporter.[12]

19 April	STS-110/Atlantis landing
25 April	Soyuz-TM 34 launch
5 May	Soyuz-TM 33 landing
5 June	STS-111/Endeavour/ISS Expedition 5 launch

---

11. "Mission Archives: STS-110," Jeanne Ryba, editor, accessed 22 February 2014, (*http://www.nasa.gov/mission_pages/shuttle/shuttlemissions/archives/sts-110.html*); *Praxis Manned Spaceflight Log: 1961–2006*, Tim Furniss and David J. Shayler, Springer, 2007, p. 709.

12. "Mission Archives: STS-110," Jeanne Ryba, editor, accessed 22 February 2014, (*http://www.nasa.gov/mission_pages/shuttle/shuttlemissions/archives/sts-110.html*); *Praxis Manned Spaceflight Log: 1961–2006*, Tim Furniss and David J. Shayler, Springer, 2007, p. 710.

9 JUNE

2002 EVA 13	**Duration:** 7:14
World EVA 222	**Spacecraft/mission:** STS-111
ESA EVA 2/U.S. EVA 126	**Crew:** Kenneth Cockrell, Paul Lockhart, Franklin Chang-Diaz, Philippe Perrin (European Space Agency)
Shuttle EVA 77	**Spacewalkers:** Franklin Chang-Diaz, Philippe Perrin
	Purpose: First of three scheduled EVAs. Install Power and Data Grapple Fixture to P6 truss; stow micrometeoroid debris shields in PMA-1; remove the Mobile Base Station's thermal blankets

On 9 June, Franklin Chang-Diaz and Philippe Perrin performed their first spacewalks ever. The STS-111 mission was Chang-Diaz' seventh and last spaceflight. At the time, he and Jerry Ross shared the record for the most spaceflights. The spacewalkers' successfully installed a Power and Data Grapple Fixture to the Station's P6 truss. The fixture allowed the later transfer of the truss to its final position. They also temporarily relocated six Service Module Debris Panels, from Endeavour's payload bay to the Station's Pressurized Mating Adapter 1 (PMA-1). The six panels are micrometeoroid debris shields. The shields were later permanently installed on the Zvezda Service Module by another crew. Chang-Diaz and Perrin also successfully completed two additional tasks: inspect and photograph the failed Control Moment Gyroscope (CMG) on the Z1 truss, and condition the Mobile Base System (MBS) to the extreme environment of space by removing its thermal blankets. The MBS was the second component of the Mobile Servicing System engineered by the Canadian Space Agency. The first component was the Canadarm2. The MBS was designed to be a mobile platform, able to carry the robotic arm and serve as a storage facility.[13]

11 JUNE

2002 EVA 14	**Duration:** 5:00
World EVA 223	**Spacecraft/mission:** STS-111
ESA EVA 3/U.S. EVA 127	**Crew:** Kenneth Cockrell, Paul Lockhart, Franklin Chang-Diaz, Philippe Perrin (European Space Agency)
Shuttle EVA 78	**Spacewalkers:** Franklin Chang-Diaz, Philippe Perrin
	Purpose: Second of three scheduled EVAs. Connect cables between Mobile Transporter and MBS; deploy Payload Orbital Replacement Unit Accommodation (POA); relocate MBS camera; secure MBS bolts

The astronauts' second EVA focused on finalizing the installation of the MBS, a movable platform. Chang-Diaz and Perrin coupled the platform's primary and backup video and data cables, as well as

13. "Mission Archives: STS-111," Jeanne Ryba, editor, accessed 25 February 2014, (*http://www.nasa.gov/mission_pages/shuttle/shuttlemissions/archives/sts-111.html*); "Mobile Base System," Canadian Space Agency, accessed 25 February 2014, (*http://www.asc-csa.gc.ca/eng/iss/mobile-base/default.asp*); *Praxis Manned Spaceflight Log: 1961–2006*, Tim Furniss and David J. Shayler, Springer, 2007, p. 715.

power connections with the Mobile Transporter. Once mounted on the Mobile Transporter, the MBS was able to move along tracks across the entire Station. The pair then attached the POA to the platform. The POA is an auxiliary grapple fixture that clutches payloads as they move along the Station's truss. The pair relocated a camera above the MBS, which provided better views of the ISS spacewalks. The astronauts finalized the installation of the MBS by securing four bolts.[14]

13 JUNE

2002 EVA 15	**Duration:** 7:16
World EVA 224	**Spacecraft/mission:** STS-111
ESA EVA 4/U.S. EVA 128	**Crew:** Kenneth Cockrell, Paul Lockhart, Franklin Chang-Diaz, Philippe Perrin (European Space Agency)
Shuttle EVA 79	**Spacewalkers:** Franklin Chang-Diaz, Philippe Perrin
	Purpose: Third of three scheduled EVAs. Replace Canadarm2 wrist roll joint

During the mission's last EVA, Chang-Diaz and Perrin repaired the Canadarm2 after a faulty joint restricted its movement. The astronauts brought the new wrist roll joint to the ISS in Endeavour's payload bay. After replacing the faulty joint, the robotic arm was restored to full use. They then aligned the new component with the wrist yaw joint, reinforced the joint and the arm with six bolts, and connected power, data, and video lines.[15]

19 June STS-111/Endeavour/ISS Expedition 4 landing

16 AUGUST

2002 EVA 16	**Duration:** 4:25
World EVA 225	**Spacecraft/mission:** International Space Station Expedition 5
Russian EVA 106/ U.S. EVA 129	**Crew:** Valery Korzun, Sergei Treschev, Peggy Whitson
Space Station EVA 116	**Spacewalkers:** Valery Korzun, Peggy Whitson
ISS EVA 11	**Purpose:** Install six micrometeoroid debris shields to Zvezda; retrieve samples from Kromka experiment

14. "Mission Archives: STS-111," Jeanne Ryba, editor, accessed 27 February 2014, (*http://www.nasa.gov/mission_pages/shuttle/shuttlemissions/archives/sts-111.html*); "Mobile Base System," Canadian Space Agency, accessed 26 February 2014, (*http://www.asc-csa.gc.ca/eng/iss/mobile-base/default.asp*); *Praxis Manned Spaceflight Log: 1961–2006*, Tim Furniss and David J. Shayler, Springer, 2007, p. 716.

15. "Mission Archives: STS-111," Jeanne Ryba, editor, accessed 27 February 2014, (*http://www.nasa.gov/mission_pages/shuttle/shuttlemissions/archives/sts-111.html*); *Praxis Manned Spaceflight Log: 1961–2006*, Tim Furniss and David J. Shayler, Springer, 2007, p. 716; "STS-111 Extravehicular Activities," Kim Dismukes, editor, accessed 28 February 2014, (*http://spaceflight.nasa.gov/shuttle/archives/sts-111/eva/index.html*).

During the crew's two spacewalks, they repaired and installed various components of the Zvezda and Zarya modules. A misconfigured oxygen valve in the Orlan spacesuits required Valery Korzun and Peggy Whitson to repeat their pre-EVA procedures, causing a 1-hour, 43-minute delay to the first EVA. Their main objective was to mount the six Service Module Debris Panels to Zvezda. To prepare for the task, the spacewalkers set up the tools and expanded the Strela boom to access the work area. Then they transported the six Service Module Debris Panels from PMA-1 and mounted them on Zvezda. The six panels are micrometeoroid debris shields intended to protect the module until its retirement. Seventeen additional shields were flown to the Station during later Shuttle missions. Due to the delay at the start of the mission, Russian flight controllers deferred the less urgent tasks of installing the Kromka experiment and collecting samples of thruster residue.[16]

26 AUGUST

2002 EVA 17	**Duration:** 5:21
World EVA 226	**Spacecraft/mission:** International Space Station Expedition 5
Russian EVA 107	**Crew:** Valery Korzun, Sergei Treschev, Peggy Whitson (NASA)
Space Station EVA 117	**Spacewalkers:** Valery Korzun, Sergei Treschev
ISS EVA 12	**Purpose:** Install attachment frame to ORU attachment frame outside FGB to Zarya exterior; replace Japan's materials experiment; mount four fairleads for EVA tethers; replace Kromka hardware

Spacewalkers Sergei Treschev and Valery Korzun were delayed due to a pressure leak in the hatch between Zvezda and Zarya. Upon exiting the Pirs hatch, the pair immediately mounted a frame that allowed equipment to be temporarily stowed on Zarya during future spacewalks. They also installed four fairleads to more efficiently route tethers during EVAs around the module. Treschev and Korzun then traded experimental devices provided by the Japanese Space Agency. The experiment required the pair to retrieve an old suitcase-sized instrument and replace it with a new panel. It measured the effects of space on engineering equipment. Because time permitted, the spacewalkers replaced the Kromka-2 deflector plate evaluator, which was an unfinished task from the EVA 10 days earlier. The cosmonauts finalized their mission with the installation of the final two amateur radio antennas on Zvezda to improve communications with ham radios on Earth.[17]

7 October STS-112/Atlantis launch

16. "*International Space Station Status Report #02-36*," John Ira Petty, editor, accessed 5 March 2014, (*http://www.nasa.gov/centers/johnson/news/station/2002/iss02-36.html*); *Praxis Manned Spaceflight Log: 1961–2006*, Tim Furniss and David J. Shayler, Springer, 2007, p. 718.

17. "International Space Station Reference: Ham Radio," Kim Dismukes, editor, accessed 8 March 2014, (*http://spaceflight.nasa.gov/station/reference/radio*); "*International Space Station Status Report #02-38*," John Ira Petty, editor, accessed 7 March 2014, (*http://www.nasa.gov/centers/johnson/news/station/2002/iss02-38.html*); *Praxis Manned Spaceflight Log: 1961–2006*, Tim Furniss and David J. Shayler, Springer, 2007, p. 718.

10 OCTOBER

2002 EVA 18	**Duration:** 7:01
World EVA 227	**Spacecraft/mission:** STS-112
U.S. EVA 130	**Crew:** Jeffrey Ashby, Pamela Melroy, David Wolf, Piers Sellers, Sandra Magnus, Fyodor Yurchikhin (Russian Space Agency)
Shuttle EVA 80	**Spacewalkers:** David Wolf, Piers Sellers
	Purpose: First of three scheduled EVAs. Connect Starboard One (S1) cables; deploy S1 antenna; install S1 camera

Mission specialists David Wolf and Piers Sellers first connected power, data, and fluid lines between the S1 and S0 trusses. They released launch bolts on a beam, to reorient the S1 radiators and allow optimum cooling. The astronauts then installed an S1 antenna, which improved voice communications with ground controllers. Upon releasing the launch restraints on CETA, Wolf and Sellers installed an external camera to the S1 truss. The manually operated CETA cart was a device capable of transporting up to 1,200 pounds (544.32 kilograms). This cart was distinguished as the "CETA-A" cart, and then later moved and renamed "CETA-B."[18]

12 OCTOBER

2002 EVA 19	**Duration:** 6:04
World EVA 228	**Spacecraft/mission:** STS-112
U.S. EVA 131	**Crew:** Jeffrey Ashby, Pamela Melroy, David Wolf, Piers Sellers, Sandra Magnus, Fyodor Yurchikhin (Russian Space Agency)
Shuttle EVA 81	**Spacewalkers:** David Wolf, Piers Sellers
	Purpose: Second of three scheduled EVAs. Install 24 Spool Positioning Devices; install camera to Destiny; release CETA launch locks

During the second of three EVAs, Wolf and Sellers installed 22 Spool Positioning Devices (SPD) on the ammonia-cooling line connections, which would prevent internal leakage from lines streaming ammonia through the Station. They also attached ammonia supply lines to the S1 radiator and mounted the ISS's second television camera on the Destiny lab. Two additional SPDs were not installed because they did not fit. At the conclusion of the EVA, the astronauts further prepared the CETA cart and other tools needed to install the next starboard truss.[19]

---

18. "International Space Station Assembly: A Construction Site in Orbit," Lyndon B. Johnson Space Center, accessed 8 March 2014, (*http://spaceflight.nasa.gov/spacenews/factsheets/pdfs/assembly.pdf*); "Mission Archives: STS-112," Jeanne Ryba, editor, accessed 8 March 2014, (*http://www.nasa.gov/mission_pages/shuttle/shuttlemissions/archives/sts-112.html*).

19. "Mission Archives: STS-112," Jeanne Ryba, editor, accessed 8 March 2014, (*http://www.nasa.gov/mission_pages/shuttle/shuttlemissions/archives/sts-112.html*); *Praxis Manned Spaceflight Log: 1961–2006*, Tim Furniss and David J. Shayler, Springer, 2007, p. 721.

14 OCTOBER

2002 EVA 20	**Duration:** 6:36
World EVA 229	**Spacecraft/mission:** STS-112
U.S. EVA 132	**Crew:** Jeffrey Ashby, Pamela Melroy, David Wolf, Piers Sellers, Sandra Magnus, Fyodor Yurchikhin (Russian Space Agency)
Shuttle EVA 82	**Spacewalkers:** David Wolf, Piers Sellers
	Purpose: Third of three scheduled EVAs. Connect lines between S0 and S1 trusses; remove S1 truss launch support

Wolf and Sellers first removed a bolt that restricted the activation of a cable cutter on the Mobile Transporter. The pair then linked ammonia lines and detached the S1 truss's launch support clamps. Ahead of schedule, they also installed SPDs on a pump motor assembly, a mechanism that circulates ammonia through the Station's cooling system. From within the Station, Sandra Magnus and Peggy Whitson, who was a member of the Expedition 5 crew, operated the Canadarm2. The robotic arm served as a work platform for the spacewalkers.[20]

18 October STS-112/Atlantis landing

30 October Soyuz-TMA 1 1aunch

10 November Soyuz-TM 34 landing

23 November STS-113/Endeavour/ISS Expedition 6 launch

26 NOVEMBER

2002 EVA 21	**Duration:** 6:45
World EVA 230	**Spacecraft/mission:** STS-113
U.S. EVA 133	**Crew:** James Wetherbee, Paul Lockhart, Michael Lopez-Alegria, John Herrington
Shuttle EVA 83	**Spacewalkers:** Michael Lopez-Alegria, John Herrington
	Purpose: First of three scheduled EVAs. Connect P1 truss cables; release CETA-B cart launch locks; install Spool Positioning Devices; install Wireless External Transceiver Assembly antenna

During the STS-113 crew's 14-day mission, they worked with the ISS Expedition 6 crew to install the Port One (P1) truss. Michael Lopez-Alegria and John Herrington of STS-113 performed the three EVAs to activate the new truss. The pair first connected power cables and fluid lines between the truss and the Station, then installed the SPDs to ensure the quick disconnection devices functioned

20. "Mission Archives: STS-112," Jeanne Ryba, editor, accessed 8 March 2014, (*http://www.nasa.gov/mission_pages/shuttle/shuttlemissions/archives/sts-112.html*); *Praxis Manned Spaceflight Log: 1961–2006*, Tim Furniss and David J. Shayler, Springer, 2007, p. 721.

correctly. The spacewalkers also disengaged launch locks on the CETA-B cart, allowing its installation during the following EVA. Lastly, the astronauts mounted the Node Wireless video system External Transceiver Assembly (WETA) antenna that provided reception from cameras on spacewalkers' helmets without a docked Shuttle.[21]

28 NOVEMBER

2002 EVA 22	**Duration:** 6:10
World EVA 231	**Spacecraft/mission:** STS-113
U.S. EVA 134	**Crew:** James Wetherbee, Paul Lockhart, Michael Lopez-Alegria, John Herrington
Shuttle EVA 84	**Spacewalkers:** Michael Lopez-Alegria, John Herrington
	Purpose: Second of three scheduled EVAs. Connect fluid lines between P1 and S0 trusses; install WETA on P1 truss; relocate CETA cart from P1 to S1 truss

On Thanksgiving Day, Lopez-Alegria and Herrington continued assembling the P1 truss. They first coupled two fluid jumpers between the P1 and S0 trusses, linking the Station's cooling system to the P1 truss. The pair retrieved the starboard keel pin and stowed it in the new truss, mounted a WETA antenna to the P1 truss, and removed the launch locks to the truss's radiator beams. Herrington rode the Canadarm2 and lifted the CETA-B cart to the S1 truss and attached it to the sister cart, which was delivered on STS-112. The astronauts concluded the EVA by rewiring an antenna installed during the previous spacewalk.[22]

30 NOVEMBER

2002 EVA 23	**Duration:** 7:00
World EVA 232	**Spacecraft/mission:** STS-113
U.S. EVA 135	**Crew:** James Wetherbee, Paul Lockhart, Michael Lopez-Alegria, John Herrington, Peggy Whitson (returned to Earth), and Valery Korzun and Sergei Treschev (Russian Space Agency, returned to Earth)
Shuttle EVA 85	**Spacewalkers:** Michael Lopez-Alegria, John Herrington
	Purpose: Third of three scheduled EVAs. Install SPDs; reconfigure MBSU cables; install NH3 tank lines; deploy UHF antenna

Lopez-Alegria and Herrington completed the installation of 33 SPDs around the Station's exterior. Herrington then disentangled an Ultra-High Frequency (UHF) communications antenna and an

21. "Mission Archives: STS-113," Jeanne Ryba, editor, accessed 7 March 2014, (*http://www.nasa.gov/mission_pages/shuttle/shuttlemissions/archives/sts-113.html*); *Praxis Manned Spaceflight Log: 1961–2006*, Tim Furniss and David J. Shayler, Springer, 2007, p. 728.

22. "Mission Archives: STS-113," Jeanne Ryba, editor, accessed 7 March 2014, (*http://www.nasa.gov/mission_pages/shuttle/shuttlemissions/archives/sts-113.html*); *Praxis Manned Spaceflight Log: 1961–2006*, Tim Furniss and David J. Shayler, Springer, 2007, p. 728.

umbilical device on the Mobile Transporter. Herrington notably finished his tasks without using the Canadarm2, which was planned to help him maneuver around the Station.[23]

7 December STS-113/Endeavour/ISS Expedition 5 landing

23. "Mission Archives: STS-113," Jeanne Ryba, editor, accessed 7 March 2014, (*http://www.nasa.gov/mission_pages/shuttle/shuttlemissions/archives/sts-113.html*); *Praxis Manned Spaceflight Log: 1961–2006*, Tim Furniss and David J. Shayler, Springer, 2007, p. 728.

2003

15 JANUARY

2003 EVA 1	**Duration:** 6:51
World EVA 233	**Spacecraft/mission:** International Space Station Expedition 6
U.S. EVA 136	**Crew:** Kenneth Bowersox, Donald Pettit, Nikolai Budarin (Russian Space Agency)
Space Station EVA 118	
ISS EVA 13	**Spacewalkers:** Kenneth Bowersox, Donald Pettit
	Purpose: Remove grit from Unity Node CBM; release 10 P1 truss radiator launch locks; install light and stanchion to CETA; measure ammonia reservoir

Nikolai Budarin was scheduled to perform the EVA on 12 December 2002 and become the first Russian spacewalker to wear an American spacesuit and use the Quest Airlock. Though Russian doctors agreed Budarin was qualified to perform Russian spacewalks, U.S. medical specialists ruled the cosmonaut physically unfit to use the U.S. spacesuits and Quest airlock. The countries have different medical requirements for use of their spacesuits and airlocks. Therefore, the EVA was delayed until 15 January 2003 and Donald Pettit replaced Budarin. The 50th spacewalk to construct and maintain the ISS was the first spacewalk for Kenneth Bowersox and Pettit. Within the Station, Nikolai Budarin choreographed the EVA and operated the television cameras on Canadarm2. At the start of the spacewalk, the crew had a few challenges. The spacewalkers had trouble opening the airlock hatch due to a strap inside the thermal hatch cover. Upon setting their spacesuits on internal battery power, Bowersox lost digital data for the suit system and was required to recycle the suit power. The astronauts then completed most of their tasks successfully. They immediately retrieved their tools and released the remaining 10 launch restraints on the P1 truss's radiator system. Mission Control in Houston then remotely unfolded the truss's center radiator to its full length of 75 feet (22.86 meters). They then prepared for the berthing of the Raffaello Multi-Purpose Logistics Module (MPLM) to the Unity Node, scheduled for March. Pettit wiped away the minor amounts of grit from the sealing ring of Unity's Common Berthing Mechanism, leaving it in pristine condition. However, the spacewalkers were unable to install a stanchion and light fixture to a CETA cart. The pair could not retrieve the stowed stanchion from the truss because a jammed pin restricted its movement. They concluded the EVA by inspecting the health of the trusses' ammonia reserves for a future Shuttle assembly flight. Upon returning to the Quest Airlock, Bowersox and Pettit cut the obstructive hatch cover strap and shut the hatch normally.[1]

16 January–1 February STS-107/Columbia (vehicle and crew were lost during re-entry)

---

1. "International Space Station Status Report #03-3," John Ira Petty, editor, accessed 15 March 2014, (*http://www.nasa.gov/centers/johnson/news/station/2003/iss03-3.html*); *Praxis Manned Spaceflight Log: 1961–2006*, Tim Furniss and David J. Shayler, Springer, 2007, p. 730.

8 APRIL

2003 EVA 2	**Duration:** 6:26
World EVA 234	**Spacecraft/mission:** International Space Station Expedition 6
U.S. EVA 137	**Crew:** Kenneth Bowersox, Donald Pettit, Nikolai Budarin (Russian Space Agency)
Space Station EVA 119	
ISS EVA 14	**Spacewalkers:** Kenneth Bowersox, Donald Pettit
	Purpose: Reconfigure S0, S1 and P1 cables; inspect the P1 truss heater; install a light and stanchion to CETA cart; reroute Control Moment Gyros cables; secured S1 Radiator Beam Valve module; replace Mobile Transporter power relay box

After the loss of the Space Shuttle Columbia on 1 February, the Expedition 6 crew performed an additional EVA to eliminate any spacewalks during the succeeding ISS crew's flight.[2] Budarin helped Bowersox and Pettit into their spacesuits, and then monitored the EVA from Destiny. First, the astronauts targeted individual tasks that were deferred or not considered urgent. Pettit installed a new power relay box in the Mobile Transporter because it had electrical glitches. Meanwhile, Bowersox rerouted the electrical connections between the S0 truss and the two adjoining trusses, S1 and P1. The reconfiguration reduced the likelihood of unintentional release of the flanking trusses. Bowersox also investigated the source of a glitch in the P1 heater, but could not find any obvious problems. After the astronauts completed their individual tasks, they met at the Z1 truss and repaired one of the Station's four orientation devices, which are collectively known as the Control Moment Gyros (CMGs). Furthermore, Bowersox and Pettit rerouted the power cables between two CMGs, preventing both from becoming disabled during a power failure. They also installed two SPDs on the fluid quick disconnect lines on Destiny's heat exchanger. Next, the spacewalkers reinstalled the S1 truss's Radiator Beam Valve module, which regulated the ammonia flow. Finally, the pair completed a task they deferred during the previous EVA and installed lights and a stanchion to the CETA cart. They were finally able to release the jammed stanchion from the truss by tapping it with a hammer.[3]

25 April	Soyuz-TMA 2/ISS Expedition 7 launch
3 May	Soyuz-TMA 1/ISS Expedition 6 landing
15–16 October	Shenzhou 5 launch and landing
18 October	Soyuz-TMA 3/ISS Expedition 8 launch
27 October	Soyuz-TMA 2/ISS Expedition 7 landing

2. Refer to *https://www.nasa.gov/columbia/home/index.html* for more details concerning the Columbia Accident.

3. "International Space Station Status Report #03-15," John Ira Petty, editor, accessed 25 March 2014, (*http://www.nasa.gov/centers/johnson/news/station/2003/iss03-15.html*); *Praxis Manned Spaceflight Log: 1961–2006*, Tim Furniss and David J. Shayler, Springer, 2007, p. 730.

2004

26 FEBRUARY

2004 EVA 1	**Duration:** 3:55
World EVA 235	**Spacecraft/mission:** International Space Station Expedition 8
Russian EVA 108/ U.S. EVA 138	**Crew:** Alexander Kaleri, Michael Foale (NASA)
	Spacewalkers: Michael Foale, Alexander Kaleri
Space Station EVA 120	**Purpose:** Replace the SKK; retrieve Micro-Particle Capturer and Space Environment Exposure Devices (MPAC/SEEDs) panel #2 and relocate panel #3; install Matryoshka experiment; replace two SKK cassettes; relocate Automated Transfer Vehicle (ATV) reflectors; install new Kromka sampling tray; reconfigure Platan-M payload experiment
ISS EVA 15	

Alexander Kaleri and Michael Foale performed the first two-person spacewalk without a crewmember occupying the Station—it was the first time the Station was unattended since November 2000. Because they were alone, the pair took a precautionary measure and temporarily installed a protective ring around the Pirs hatch, preventing snagging when entering and leaving the airlock. Their objective was to install European and Japanese scientific packages and retrieve old data on the Zvezda module. They replaced two of the Russian SKK, which contained materials measuring the effects of the space environment. Next, the spacewalkers retrieved a panel and relocated another panel of the Japanese-made MPAC/SEED. The experiment studied micrometeoroid impacts. Kaleri and Foale then installed a new Russian experiment, Matryoshka, a torso-like device comprised of simulative human tissue. The data was used to study the effects of radiation on long-term crewmembers. The EVA was scheduled for about 5 hours, 30 minutes, but was cut short when Kaleri felt unusually warm and noticed water drops in his helmet visor. Soon after, Russian flight controllers identified a cooling and dehumidifying failure in the Russian Orlan suit. Both spacewalkers immediately returned to the Pirs module and repressurized the airlock. Foale quickly located and straightened out a kink in the liquid cooling garment umbilical, allowing water to flow normally again. Work on the ATV reflectors, Kromka thrusters, and the Platan-M payload were deferred.[1]

18 April Soyuz-TMA 4/ISS Expedition 9 launch

29 April Soyuz-TMA 3/ISS Expedition 8 landing

1. "International Space Station Status Report #04-11," John Ira Petty, editor, accessed 28 March 2014, (*http://www.nasa.gov/centers/johnson/news/station/2004/iss04-11.html*); *Praxis Manned Spaceflight Log: 1961–2006*, Tim Furniss and David J. Shayler, Springer, 2007, p. 730.

24 JUNE

2004 EVA 2	**Duration:** 0:14
World EVA 236	**Spacecraft/mission:** International Space Station Expedition 9
Russian EVA 109/ U.S. EVA 139	**Crew:** Gennady Padalka, Michael Fincke (NASA)
	Spacewalkers: Gennady Padalka, Michael Fincke
Space Station EVA 121	**Purpose:** Replace malfunctioning circuit breaker
ISS EVA 16	

For over a month, the crew repaired the faulty Orlan spacesuit and serviced the Station in preparation for the EVA. However, at the start of the spacewalk, Russian flight controllers noticed Michael Fincke's primary oxygen bottle depressurizing too quickly. Within 14 minutes, the spacewalkers abandoned the EVA. Fincke was entirely stable and did not require the reserved oxygen tank. They later identified the oxygen control lever as the cause of the depressurization. The crew later completed the mission on 30 June.[2]

30 JUNE

2004 EVA 3	**Duration:** 5:40
World EVA 237	**Spacecraft/mission:** International Space Station Expedition 9
Russian EVA 110/ U.S. EVA 140	**Crew:** Gennady Padalka, Michael Fincke (NASA)
	Spacewalkers: Gennady Padalka, Michael Fincke
Space Station EVA 122	**Purpose:** Replace Remote Power Control Module (RPCM) on a CMG
ISS EVA 17	

The spacewalkers and Russian and U.S. mission controllers communicated effortlessly in a well-choreographed and seamless EVA. Unlike the previous EVA on 24 June that was terminated early due to a spacesuit problem, the 30 June 30 spacewalk was completed smoothly, even allowing Padalka and Fincke to complete additional tasks. Their primary and most urgent job was completed an hour ahead of schedule. The pair replaced a faulty RPCM and restored power to a CMG, one of the Station's orientation devices. With time permitting, Padalka and Fincke additionally installed two flexible fabric handrails, mounted a Kromka contamination monitor to the thruster exhausts, and attached two handrail end caps on the airlock.[3]

2. "International Space Station Status Report #04-32," John Ira Petty, editor, accessed 28 March 2014, (*http://www.nasa.gov/centers/johnson/news/station/2004/iss04-32.html*); *Praxis Manned Spaceflight Log: 1961–2006*, Tim Furniss and David J. Shayler, Springer, 2007, p. 745.

3. "International Space Station Status Report #04-32," John Ira Petty, editor, accessed 1 April 2014, (*http://www.nasa.gov/centers/johnson/news/station/2004/iss04-36.html*); *Praxis Manned Spaceflight Log: 1961–2006*, Tim Furniss and David J. Shayler, Springer, 2007, p. 746.

3 AUGUST

2004 EVA 4	**Duration:** 4:30
World EVA 238	**Spacecraft/mission:** International Space Station Expedition 9
Russian EVA 111/ U.S. EVA 141	**Crew:** Gennady Padalka, Michael Fincke (NASA)
	Spacewalkers: Gennady Padalka, Michael Fincke
Space Station EVA 123	**Purpose:** Install two ATV antennas; remove six ATV laser reflectors and install four new reflectors; uncouple cable from ATV camera; replace Kromka, SKK, and Platan-M experiments
ISS EVA 18	

To prepare for the docking of the first ATV, Padalka and Fincke installed the corresponding and necessary hardware. The ATV, known as "Jules Verne," was an unpiloted cargo ship designed by the European Space Agency (ESA) with the capacity to carry seven tons of equipment and supplies. Furthermore, the ATV boosted the Station's orbit and carried away waste in its atmospheric burn up. Before they completed docking preparations, the spacewalkers first replaced the SKK cassette and the Kromka thruster experiment, as well as retrieved the Platan-M data. They then installed two ATV antennas and docking equipment, and replaced six laser reflectors with four more advanced counterparts. Finally, they disconnected the cable of a failed ATV camera.[4]

3 SEPTEMBER

2004 EVA 5	**Duration:** 5:21
World EVA 239	**Spacecraft/mission:** International Space Station Expedition 9
Russian EVA 112/ U.S. EVA 142	**Crew:** Gennady Padalka, Michael Fincke (NASA)
	Spacewalkers: Gennady Padalka, Michael Fincke
Space Station EVA 124	**Purpose:** Replace Zarya thermal system fluid control unit; mount four tether fairleads on Zarya; install three ATV antennas and remove five antenna covers; install four handrails on Pirs hatch; photograph MPAC/SEED
ISS EVA 19	

During the crew's fourth and final EVA, they continued preparations to dock the first ESA-provided ATV. Padalka and Fincke first replaced the Zarya module's thermal system fluid control unit, a device that measured coolant levels. Next, they mounted tether guides on four Zarya handrails to prevent snagged tethers. The spacewalkers took a break during orbital darkness and updated the Houston flight controllers on the ISS's orientation. The data helped determine if the Orlan suits affected the Station's position. The pair then installed three communication antennas on Zvezda, which later guided the docking ATV. They attached protective handrail covers to Pirs' airlock hatches, to avoid inadvertent

4. "International Space Station Status Report #04-43," John Ira Petty, editor, accessed 5 April 2014, (*http://www.nasa.gov/centers/johnson/news/station/2004/iss04-43.html*); *Praxis Manned Spaceflight Log: 1961–2006*, Tim Furniss and David J. Shayler, Springer, 2007, p. 746.

snags between the tethers and handrails. After Fincke photographed the MPAC/SEED experiment, they concluded the EVA.[5]

13 October Soyuz-TMA 5/ISS Expedition 10 launch

23 October Soyuz-TMA 4/ISS Expedition 9 landing

5. "International Space Station Status Report #04-50," John Ira Petty, editor, accessed 5 April, 2014, (*http://www.nasa.gov/centers/johnson/news/station/2004/iss04-50.html*); *Praxis Manned Spaceflight Log: 1961–2006*, Tim Furniss and David J. Shayler, Springer, 2007, p. 746.

2005

26 JANUARY

2005 EVA 1	**Duration:** 5:28
World EVA 240	**Spacecraft/mission:** International Space Station Expedition 10
Russian EVA 113/ U.S. EVA 143	**Crew:** Salizhan Sharipov (Russian Space Agency), Leroy Chiao
	Spacewalkers: Salizhan Sharipov, Leroy Chiao
Space Station EVA 125	**Purpose:** Install the Universal Work Platform; install Rokviss robotic arm and antenna; relocate MPAC/SEED experiment; photograph contamination on Zvezda vent; install Biorisk experiment
ISS EVA 20	

During their first EVA together, Chiao and Sharipov installed the Universal Work Platform and numerous experiments to Zvezda's exterior. Above the platform, the crew attached the German experiment Rokviss, which consisted of a small flexible robotic arm, an illumination device, and a power source. The installation of the Rokviss antenna allowed both the ISS crew and German operators to control the arm remotely. The crew then transported the MPAC/SEED experiment to an adjacent bracket on Zvezda. Sharipov also photographed white and brown residues and an oily substance around the Zvezda vent. The deposits were byproducts of the Elektron oxygen generator, Vozdukh carbon dioxide scrubber, and a particle purification device. The crew concluded the spacewalk with the installation of the Russian experiment Biorisk. The study observed the effects of the space environment on microorganisms.[1]

28 MARCH

2005 EVA 2	**Duration:** 4:30
World EVA 241	**Spacecraft/mission:** International Space Station Expedition 10
Russian EVA 114/ U.S. EVA 144	**Crew:** Leroy Chiao, Salizhan Sharipov (Russian Space Agency)
	Spacewalkers: Salizhan Sharipov, Leroy Chiao
Space Station EVA 126	**Purpose:** Install three ATV antennas; jettison nanosatellite experiment; install ATV GPS; observe and photograph ATV antennas and laser reflector
ISS EVA 21	

Chiao and Sharipov continued preparations for the first ATV docking and deployed a small experimental satellite. The spacewalkers first installed three communication antennas on Zvezda. The antennas are a segment of the Proximity Communications Equipment (PCE) that provided interaction between the ATV and service module during docking. Two hours later, from a ladder on Pirs, Sharipov

1. "International Space Station Status Report #05-04," John Ira Petty, editor, accessed 5 April 2014, (*http://www.nasa.gov/centers/johnson/news/station/iss05-04.html*); *Praxis Manned Spaceflight Log: 1961–2006*, Tim Furniss and David J. Shayler, Springer, 2007, p. 749.

manually released the 1-foot, (0.30-meter), 11-pound (4.99-kilogram) German nanosatellite into orbit. Chiao photographed the deployment. The small satellite tested new and untried satellite techniques, operations, and system sensors. Two hours later, Russian controllers confirmed a good signal from the nanosatellite. Also, with the assistance of Russian flight controllers, the crew installed another part of the PCE, called the Global Positioning System (GPS) receiver. The device provided approaching ATVs the Station's position. In addition, the pair examined and photographed ATV components to confirm their locations with Russian controllers. As they returned to the Pirs airlock, the spacewalkers secured cabling on Zvezda. Due to a circuit breaker failure, one of three CMGs stopped functioning. The two functioning CMGs sustained the ISS's orientation until the end of the EVA. The thrusters could not be used during the spacewalk due to its proximity to the work area.[2]

14 April Soyuz-TMA 6/ISS Expedition 11 launch

24 April Soyuz-TMA 5/ISS Expedition 10 landing

26 July STS-114/Discovery launch

30 JULY

2005 EVA 3	**Duration:** 6:50
World EVA 242	**Spacecraft/mission:** STS-114
Japanese EVA 3/ U.S. EVA 145	**Crew:** Eileen Collins, James Kelly, Charles Camarda, Wendy Lawrence, Stephen Robinson, Andrew Thomas, Soichi Noguchi (JAXA)
Shuttle EVA 86	**Spacewalkers:** Soichi Noguchi, Stephen Robinson
	Purpose: First of three scheduled EVAs. Demonstrate repair techniques to the Shuttle's Thermal Protection System using Emittance Wash Applicator on broken tile test article; install stowage platform base and cables; reroute CMG-2 power cables; replace GPS antenna

Mission specialist Soichi Noguchi was the second JAXA astronaut to perform a spacewalk. The occasion marked the third Japanese EVA. With Discovery docked on the ISS, Noguchi and Stephen Robinson executed the first of three EVAs. They first demonstrated repairing deliberately damaged tiles to test the Emittance Wash Applicator, a new tile repair tool for the Shuttle Thermal Protection System. The spacewalkers also furthered the ISS construction and installed the External Stowage Platform Attachment Device, which consisted of the platform base and cables. They also replaced the GPS antenna and rerouted power to the CMG-2.[3]

2. "International Space Station Status Report #05-16," John Ira Petty, editor, accessed 5 April 2014, (*http://www.nasa.gov/centers/johnson/news/station/iss05-16.html*); *Praxis Manned Spaceflight Log: 1961–2006*, Tim Furniss and David J. Shayler, Springer, 2007, p. 750.

3. "Mission Archives: STS-114," Jeanne Ryba, editor, accessed 2 April 2014, (*http://www.nasa.gov/mission_pages/shuttle/shuttlemissions/archives/sts-114.html*); *Praxis Manned Spaceflight Log: 1961–2006*, Tim Furniss and David J. Shayler, Springer, 2007, p. 756.

1 AUGUST

2005 EVA 4	**Duration:** 7:14
World EVA 243	**Spacecraft/mission:** STS-114
Japanese EVA 4/ U.S. EVA 146	**Crew:** Eileen Collins, James Kelly, Charles Camarda, Wendy Lawrence, Stephen Robinson, Andrew Thomas, Soichi Noguchi (JAXA)
Shuttle EVA 87	**Spacewalkers:** Soichi Noguchi, Stephen Robinson
	Purpose: Second of three scheduled EVAs. Replace CMG-1; set up tools for next EVA

During the STS-114 crew's second EVA, Robinson and Noguchi disconnected and stowed the failed CMG-1. They retrieved the replacement CMG from the payload bay. Its installation restored the ISS to four, properly functioning CMGs. The astronauts prepared tools for the following EVA and concluded the excursion.[4]

3 AUGUST

2005 EVA 5	**Duration:** 6:01
World EVA 244	**Spacecraft/mission:** STS-114
Japanese EVA 5/ U.S. EVA 147	**Crew:** Eileen Collins, James Kelly, Charles Camarda, Wendy Lawrence, Stephen Robinson, Andrew Thomas, Soichi Noguchi (JAXA)
Shuttle EVA 88	**Spacewalkers:** Soichi Noguchi, Stephen Robinson
	Purpose: Third of three scheduled EVAs. Inspect and remove two protruding Shuttle tile gap fillers; install the external stowage platform to ISS; deploy the Materials International Space Station Experiment (MISSE 5)

Robinson rode the Canadarm2 to Discovery's underside and became the first astronaut to venture there and inspect and repair the Shuttle during a spacewalk. With his gloved hand, Robinson gently tugged and retrieved two protruding gap fillers between the orbiter's thermal protection tiles. The job was expected to be difficult, but he completed this unplanned repair task easily. The crew also completed the installation of the external stowage platform, which later stored spare parts from future Station EVAs. Robinson and Noguchi installed the MISSE. This experiment exposed materials to the harsh space environments for months.[5]

9 August STS-114/Discovery landing

---

4. "Mission Archives: STS-114," Jeanne Ryba, editor, accessed 2 April 2014, (*http://www.nasa.gov/mission_pages/shuttle/shuttlemissions/archives/sts-114.html*); *Praxis Manned Spaceflight Log: 1961–2006*, Tim Furniss and David J. Shayler, Springer, 2007, p. 756.

5. "Mission Archives: STS-114," Jeanne Ryba, editor, accessed 2 April 2014, (*http://www.nasa.gov/mission_pages/shuttle/shuttlemissions/archives/sts-114.html*); *Praxis Manned Spaceflight Log: 1961–2006*, Tim Furniss and David J. Shayler, Springer, 2007, p. 756.

18 AUGUST

2005 EVA 6	**Duration:** 4:58
World EVA 245	**Spacecraft/mission:** International Space Station Expedition 11
Russian EVA 115/ U.S. EVA 148	**Crew:** Sergei Krikalev (Russian Space Agency), John Phillips (NASA), Roberto Vittori (European Space Agency)
Space Station 127	**Spacewalkers:** Sergei Krikalev, John Phillips
ISS EVA 22	**Purpose:** Retrieve Biorisk experiment canister, MPAC/SEED panel number three, and Matryoshka experiment; install a reserve camera on Zvezda; photograph Korma experiment

During John Phillips' first spacewalk, he and veteran spacewalker Sergei Krikalev retrieved samples from various experiments, installed a spare television camera for ATV dockings, and photographed the Korma experiment. They retrieved a canister from the Biorisk experiment, a MPAC/SEED panel, and the simulative torso-shaped Matryoshka. The collected data were later studied in the Station and on Earth. The task to relocate a grapple fixture, from the Strela crane to PMA-3, was deferred to a future spacewalk.[6]

30 September Soyuz-TMA 7/ISS Expedition 12 launch

10 October Soyuz-TMA 6/ISS Expedition 11 landing

12–16 October Shenzhou 6 launch and landing

7 NOVEMBER

2005 EVA 7	**Duration:** 5:22
World EVA 246	**Spacecraft/mission:** International Space Station Expedition 12
Russian EVA 116/ U.S. EVA 149	**Crew:** William McArthur, Valeri Tokarev (Russian Space Agency)
	Spacewalkers: William McArthur, Valeri Tokarev
Space Station EVA 128	**Purpose:** Install TV camera to P1 truss; replace S1 rotary joint motor controller; jettison Floating Potential Probe; replace the failed circuit breaker of the Mobile Transporter
ISS EVA 23	

For the first time since 8 April 2003, an ISS crew wore the U.S. spacesuits and exited the Quest Airlock. Furthermore, Tokarev became the first Russian spacewalker to use U.S. facilities in the Station. The pair required 2 hours and 10 minutes to install a camera on the P1 truss. The camera later greatly assisted in the installation of the Port Three (P3) and Port Four (P4) trusses. While waiting for the required sunlight to jettison the failed Floating Potential Probe (FPP), the spacewalkers moved on to

6. "Expedition 11: Station Crew Completes Spacewalk," John Ira Petty, editor, accessed 5 April 2014, (*http://www.nasa.gov/mission_pages/station/expeditions/expedition11/exp11_spacewalk.html*); *Praxis Manned Spaceflight Log: 1961–2006*, Tim Furniss and David J. Shayler, Springer, 2007, p. 753.

a task scheduled for a future EVA and retrieved a failed rotary joint motor controller. Tokarev and McArthur then successfully jettisoned the FPP into orbit, where it later burned up during atmospheric reentry. Afterward, they completed an unscheduled task and replaced a failed circuit breaker in the Mobile Transporter.[7]

7. "International Space Station: Spacewalkers Install New Camera Assembly, Jettison FPP," John Ira Petty, editor, accessed 5 April 2014, (*http://www.nasa.gov/mission_pages/station/expeditions/expedition12/exp12_eva.html*); *Praxis Manned Spaceflight Log: 1961–2006*, Tim Furniss and David J. Shayler, Springer, 2007, p. 758.

2006

3 FEBRUARY

2006 EVA 1	**Duration:** 5:43
World EVA 247	**Spacecraft/mission:** International Space Station Expedition 12
Russian EVA 117/ U.S. EVA 150	**Crew:** William McArthur, Valeri Tokarev (Russian Space Agency)
Space Station EVA 129	**Spacewalkers:** William McArthur, Valeri Tokarev
ISS EVA 24	**Purpose:** Jettison SuitSat, retired Orlan suit with ham radio and batteries; relocate Strela boom adapter to PMA-3; secure Mobile Transporter cable; retrieve Biorisk experiment; photograph exterior of service module

The Expedition 12 crew's second spacewalk was postponed from December 2005 to February 2006, to allow more time for preparations. McArthur and Tokarev wore the Orlan M suits and exited from the Pirs airlock. They manually jettisoned a retired Orlan spacesuit into orbit that had been repurposed and equipped with three batteries, a ham radio, and internal sensors. For a few days, the orbiting "SuitSat" emitted children's voices and salutations in six languages to amateur radio operators. Within a few weeks, SuitSat burned up during reentry. The spacewalkers also relocated a Strela grapple fixture from the Zarya module to PMA-3, to create a temporary storage area for the Zvezda debris panels. McArthur and Tokarev then relocated and secured a mobile transporter cable, which was inadvertently severed on 16 December 2005. They also retrieved the Biorisk experiment and photographed the Russian Micrometeoroid Measuring System, the exterior of Zvezda, a ham radio antenna, and a fuel drain outlet pipe.[1]

29 March Soyuz-TMA 8/ISS Expedition 13 launch

8 April Soyuz-TMA 7/ISS Expedition 12 landing

1. "International Space Station: Spacewalkers Install New Camera Assembly, Jettison FPP," John Ira Petty, editor, accessed 5 April 2014, (*http://www.nasa.gov/mission_pages/station/expeditions/expedition12/exp12_eva.html*); *Praxis Manned Spaceflight Log: 1961–2006*, Tim Furniss and David J. Shayler, Springer, 2007, p. 758.

FIGURE 6. **SuitSat.** Backdropped by the blackness of space and Earth's horizon, a spacesuit-turned-satellite called SuitSat began its orbit around Earth after it was released by the Expedition 12 crewmembers during an EVA on 3 February 2006. SuitSat, a Russian Orlan spacesuit whose utility had expired, was outfitted by the crew with three batteries, internal sensors and a radio transmitter, which faintly transmitted recorded voices of school children to amateur radio operators worldwide. (NASA ISS012-E-16905)

1 JUNE

2006 EVA 2	**Duration:** 6:31
World EVA 248	**Spacecraft/mission:** International Space Station Expedition 13
Russian EVA 118/ U.S. EVA 151	**Crew:** Pavel Vinogradov (Russian Space Agency), Jeffrey Williams (NASA)
	Spacewalkers: Pavel Vinogradov, Jeffrey Williams
Space Station EVA 130	**Purpose:** Install new nozzle to Elektron hydrogen exhaust; retrieve Kromka and Biorisk experiments; relocate ATV antenna cable; replace MBS camera
ISS EVA 25	

During the crew's first of two EVAs, Vinogradov and Williams repaired various experiments on U.S. and Russian segments. Wearing Orlan spacesuits, they rerouted the hydrogen vent shared by the oxygen generator Elektron and the carbon dioxide eliminator Vozdukh. The Elektron unit divided water into oxygen and hydrogen and released the latter through a vent and into space. By installing a valve nozzle to the hydrogen exhaust, the spacewalkers minimized the buildup in the vent created by the hydrogen. Vinogradov then retrieved samples from the Kromka experiment, and then relocated the

ATV antenna cable to improve system performance. Meanwhile, Williams removed a Biorisk experiment canister and another contamination experimental device. Their final completed task to replace a faulty camera on the MBS required the spacewalkers to extend the maximum EVA time.[2]

4 July STS-121/Discovery launch

8 JULY

2006 EVA 3	**Duration:** 7:31
World EVA 249	**Spacecraft/mission:** STS-121
U.S. EVA 152	**Crew:** Steven Lindsey, Mark Kelly, Stephanie Wilson, Michael Fossum, Piers Sellers, Lisa Nowak, Thomas Reiter (European Space Agency)
Shuttle EVA 89	**Spacewalkers:** Piers Sellers, Michael Fossum
	Purpose: First of three scheduled EVAs. Install blade blocker to umbilical system cables; reroute cables through Interface Umbilical Assembly and protect cables; evaluate the use of Canadarm2 and Orbital Boom Sensor System as an EVA work platform

A function of the S0 truss's Trailing Umbilical System is to sever certain cables; however, that component inadvertently cut its own power and data cables. To protect undamaged electrical lines, Michael Fossum and Piers Sellers installed a blade blocker and rerouted the cables through the Interface Umbilical Assembly. The cable reconfiguration resulted in the replacement of the Trailing Umbilical System with the Mobile Transporter. The spacewalkers then tested the simultaneous use of the Canadarm2 and 50-foot-long (15-meter) Orbital Boom Sensor System, which together served as an EVA work platform for thermal protection repairs on the Shuttle and for accessing difficult areas on the Station.[3]

10 JULY

2006 EVA 4	**Duration:** 6:47
World EVA 250	**Spacecraft/mission:** STS-121
U.S. EVA 153	**Crew:** Steven Lindsey, Mark Kelly, Stephanie Wilson, Michael Fossum, Piers Sellers, Lisa Nowak, Thomas Reiter (European Space Agency)
Shuttle EVA 90	**Spacewalkers:** Piers Sellers, Michael Fossum
	Purpose: Second of three scheduled EVAs. Restore Mobile Transporter to full operation; replace Interface Umbilical Assembly

---

2. "International Space Station: Station Crew Winds up Successful Spacewalk," Amiko Kauderer, editor, accessed 22 March 2014, (*http://www.nasa.gov/mission_pages/station/expeditions/expedition13/eva1.html*); *Praxis Manned Spaceflight Log: 1961–2006*, Tim Furniss and David J. Shayler, Springer, 2007, p. 763.

3. "Mission Archives: STS-121," Jeanne Ryba, editor, accessed 20 March 2014, (*http://www.nasa.gov/mission_pages/shuttle/shuttlemissions/archives/sts-121.html*); *Praxis Manned Spaceflight Log: 1961–2006*, Tim Furniss and David J. Shayler, Springer, 2007, p. 767.

After the rerouting of the Mobile Transporter cable during the previous EVA, Fossum and Sellers restored the Canadarm2 railcar to full operation. The astronauts also installed a new Interface Umbilical Assembly that, unlike its predecessor, did not include a blade. During the excursion, Seller secured Fossum's SAFER latches that unexpectedly loosened.[4]

12 JULY

2006 EVA 5	**Duration:** 7:11
World EVA 251	**Spacecraft/mission:** STS-121
U.S. EVA 154	**Crew:** Steven Lindsey, Mark Kelly, Stephanie Wilson, Michael Fossum, Piers Sellers, Lisa Nowak, Thomas Reiter (European Space Agency)
Shuttle EVA 91	**Spacewalkers:** Piers Sellers, Michael Fossum,
	Purpose: Third of three scheduled EVAs. Test and photograph Non-Oxide Adhesive eXperiment (NOAX) repairs on sample tiles; photograph Discovery's port wing; relocate grapple from integrated cargo carrier to S1 truss's ammonia tank

The primary objective of the last STS-121 spacewalk was testing repairs on the Thermal Protection System Reinforced Carbon-Carbon panels. Fossum and Sellers tested NOAX, a pre-ceramic sealant combined with silicon carbide powder, on two cracks and three gouge repairs. Piers inadvertently lost one of the NOAX spatulas used in the repair demonstration. They photographed the repaired tiles and Discovery's port wing. Additionally, the spacewalkers had enough time to transfer a fixed grapple from Discovery's payload bay to the S1 truss's ammonia tank. The extra task assisted the transfer of the tank during a future excursion.[5]

17 July STS-121/Discovery landing

3 AUGUST

2006 EVA 6	**Duration:** 5:54
World EVA 252	**Spacecraft/mission:** International Space Station Expedition 13
ESA EVA 5 /U.S. EVA 155	**Crew:** Pavel Vinogradov (Russian Space Agency), Jeffrey Williams (NASA), Thomas Reiter (European Space Agency)
Space Station EVA 131	**Spacewalkers:** Jeffrey Williams, Thomas Reiter
ISS EVA 26	**Purpose:** Install Floating Potential Measurement Unit; install third and fourth MISSE; install three SPDs; install two jumpers on S1 truss; replace S1 computer; test infrared camera

4. Ibid.

5. "Mission Archives: STS-121," Jeanne Ryba, editor, accessed 20 March 2014, (*http://www.nasa.gov/mission_pages/shuttle/shuttlemissions/archives/sts-121.html*); *Praxis Manned Spaceflight Log: 1961–2006*, Tim Furniss and David J. Shayler, Springer, 2007, p. 767.

With the subsequent arrival of European astronaut Thomas Reiter, the ISS housed a three-man crew for the first time since the Columbia accident. Experienced spacewalker Vinogradov aided Williams and Reiter through the preparatory EVA procedures. Astronaut Stephen Bowen in Houston Mission Control coached the spacewalkers as they installed the Floating Potential Measurement Unit (FPMU), a tool that gauged the Station's electrical potential. Afterward, they installed the third and fourth MISSE containers. The spacewalkers then individually assumed separate assignments. Williams installed a controller to the S1 truss's rotary joint while Reiter replaced an S1 computer. Still on the truss, Williams installed an SPD and starboard jumper to improve the flow of ammonia. Reiter installed two more SPDs: one on the radiator beam module and the other by the new port jumper. Next, the spacewalkers worked together and tested an infrared camera, designed to locate thermal protection damage. Furthermore, Williams completed two supplementary tasks: install a light on the CETA cart and replace a faulty GPS antenna. Meanwhile, Reiter assembled a vacuum system valve on the Destiny lab for future studies. To efficiently maximize the time remaining, Mission Control assigned the spacewalkers additional tasks. The pair relocated two portable foot restraints, photographed a scratch on the airlock, and retrieved a ball stack restraint that connected PMA-1 hardware to the Station.[6]

9 September STS-115/Atlantis launch

12 SEPTEMBER

2006 EVA 7	**Duration:** 6:26
World EVA 253	**Spacecraft/mission:** STS-115
U.S. EVA 156	**Crew:** Brent Jett, Jr.; Christopher Ferguson; Heidemarie Stefanyshyn-Piper; Joseph Tanner; Daniel Burbank; Steven MacLean (Canadian Space Agency)
Shuttle EVA 92	
	Spacewalkers: Joseph Tanner, Heidemarie Stefanyshyn-Piper
	Purpose: First of three scheduled EVAs. Connect P3 and P4 power cables; remove launch restraints on solar array components

The night prior to the first STS-115 spacewalk, Heidemarie Stefanyshyn-Piper and Joseph Tanner slept in the Quest Airlock as part of the prebreathe protocol to eliminate nitrogen in their bloodstreams and mitigate the possibility of Decompression Sickness. The campouts in the Quest airlock shortened the preparatory EVA procedures the next day. The objectives of the three STS-115 spacewalks were to deploy solar arrays and radiators and install the P3 and P4 trusses. Stefanyshyn-Piper and Tanner first coupled the trusses' umbilical power cables. Then, they shifted their focus to the solar array assembly. The pair detached the launch restraints on the new solar array blanket box, the beta gimbal assembly, and the solar array wings. To prepare for the STS-116 mission, Stefanyshyn-Piper and Tanner removed

6. "International Space Station: Station Crewman Back Inside After Spacewalk," Amiko Kauderer, editor, accessed 22 March 2014, (*http://www.nasa.gov/mission_pages/station/expeditions/expedition13/exp13_eva_08_03_2006.html*); *Praxis Manned Spaceflight Log: 1961–2006*, Tim Furniss and David J. Shayler, Springer, 2007, p. 764.

two circuit interrupt devices. Ahead of schedule, they worked on the following day's spacewalk and began to set up the Solar Alpha Rotary Joint (SARJ), a component of the arrays that tracked the Sun.[7]

13 SEPTEMBER

2006 EVA 8	**Duration:** 7:11
World EVA 254	**Spacecraft/mission:** STS-115
Canadian EVA 3/ U.S. EVA 157	**Crew:** Brent Jett, Jr.; Christopher Ferguson; Heidemarie Stefanyshyn-Piper; Joseph Tanner; Daniel Burbank; Steven MacLean (Canadian Space Agency)
Shuttle EVA 93	**Spacewalkers:** Daniel Burbank, Steven MacLean
	Purpose: Second of three scheduled EVAs. Release launch restraints from the SARJ

Together, Steven MacLean and Daniel Burbank completed their first spacewalk ever. The excursion also marked the third Canadian EVA. Though they successfully completed their primary task and unlocked the SARJ, the pair faced many minor problems: a malfunctioning helmet, a broken socket tool, a jammed bolt that required effort from both astronauts, and another bolt loosened by the device constructed to secure it.[8]

15 SEPTEMBER

2006 EVA 9	**Duration:** 6:42
World EVA 255	**Spacecraft/mission:** STS-115
U.S. EVA 158	**Crew:** Brent Jett, Jr.; Christopher Ferguson; Heidemarie Stefanyshyn-Piper; Joseph Tanner; Daniel Burbank; Steven MacLean (Canadian Space Agency)
Shuttle EVA 94	**Spacewalkers:** Joseph Tanner, Heidemarie Stefanyshyn-Piper
	Purpose: Third of three scheduled EVAs. Power cooling radiator for solar arrays; replace S-band antenna

Stefanyshyn-Piper and Tanner activated the solar arrays' cooling radiator for the first time and replaced an S-band radio antenna that provided standby communication between the ISS and ground control. Furthermore, they completed tasks scheduled for future EVAs; they installed a protective insulation around another communication device and tested an infrared camera by photographing the Shuttle's wings.[9]

---

7. "Mission Archives: STS-115," Jeanne Ryba, editor, accessed 21 March 2014, (*http://www.nasa.gov/mission_pages/shuttle/shuttlemissions/archives/sts-115.html*); *Praxis Manned Spaceflight Log: 1961–2006*, Tim Furniss and David J. Shayler, Springer, 2007, p. 770.

8. Ibid.

9. Ibid.

17 September Soyuz-TMA 9/ISS Expedition 14 launch

21 September STS-115/Atlantis landing

28 September Soyuz-TMA 8/ISS Expedition 13 landing

22 NOVEMBER

2006 EVA 10	**Duration:** 5:38
World EVA 256	**Spacecraft/mission:** International Space Station Expedition 14
Russian EVA 119/ U.S. EVA 159	**Crew:** Michael Lopez-Alegria (NASA), Mikhail Tyurin (Russian Space Agency), Thomas Reiter (European Space Agency)
Space Station EVA 132	**Spacewalkers:** Michael Tyurin, Michael Lopez-Alegria
ISS EVA 27	**Purpose:** Commercial space task of hitting a golf ball; stow Kurs antenna; relocate ATV antenna; install BTN-Neutron-Radiation-ISS experiment

Mikhail Tyurin was designated the lead spacewalker during the first of five Expedition 14 EVAs. As a commercial space marketing venture, a Canadian golf company funded the spacewalkers' first chief assignment, to golf in space. Michael Lopez-Alegria positioned the tee and secured Tyurin's feet, as the latter volleyed a ball into orbit with a one-hand shot. Unlike a standard 1.62-ounce (45.92-gram) golf ball, it weighed only a tenth of an ounce (2.83 grams). The pair then inspected the Kurs antenna, which refused to retract into the Progress 23 vehicle as it docked on 26 October. Both the spacewalkers and the Russian flight controllers could not stow the stubborn antenna and therefore abandoned the task. Lopez-Alegria and Tyurin then relocated a WAL antenna because it obstructed a Zvezda booster engine. The WAL antenna is designed to guide ATVs during dockings. The spacewalkers also installed the BTN-Neutron-Radiation-ISS experiment, which measures the effect of solar flares on the Station's exterior. They deferred the final task, an inspection of bolts on a Strela crane.[10]

9 December STS-116/Discovery launch

---

10. "International Space Station: Spacewalkers Tee Off on Science, Mechanics," Amiko Kauderer, editor, accessed 22 March 2014, (*http://www.nasa.gov/mission_pages/station/expeditions/expedition14/exp14_eva_112206.html*).

12 DECEMBER

2006 EVA 11	**Duration:** 6:36
World EVA 257	**Spacecraft/mission:** STS-116
ESA EVA 6/U.S. EVA 160	**Crew:** Mark Polansky, William Oefelein, Sunita Williams, Joan Higginbotham, Nicholas Patrick, Robert Curbeam, Thomas Reiter (European Space Agency), Christer Fuglesang (European Space Agency)
Shuttle EVA 95	**Spacewalkers:** Robert Curbeam, Christer Fuglesang
	Purpose: First of four scheduled EVAs. Remove launch locks from the P5 truss; aid alignment and secure the P5 truss to the P4 truss; relocate radiator grapple fixture; replace failed camera on S1 truss

This was the first of a series of four EVAs performed by the Discovery crew. Their primary objectives were to install the Port Five (P5) truss and to redirect the Station's wiring to the permanent power grid. On the day before the first EVA, Nicholas Patrick used Discovery's robotic arm to remove the P5 truss from the Shuttle's payload bay and pass it to ISS's arm. Curbeam and ESA astronaut Christer Fuglesang then aligned the truss, attached it to the end of the P3 and P4 trusses, and plugged it into the existing truss structure's power. Curbeam and Fuglesang then carried out the additional tasks of replacing a failed camera that was necessary for future Station assemblies. The pair concluded their spacewalk by removing the launch restraints on the P5 truss, as well as opening a latch on the end of the truss to attach the Port Six (P6) truss permanently on a later EVA.[11]

14 DECEMBER

2006 EVA 12	**Duration:** 5:00
World EVA 258	**Spacecraft/mission:** STS-116
ESA EVA 7/U.S. EVA 161	**Crew:** Mark Polansky, William Oefelein, Sunita Williams, Joan Higginbotham, Nicholas Patrick, Robert Curbeam, Thomas Reiter (European Space Agency), Christer Fuglesang (European Space Agency)
Shuttle EVA 96	**Spacewalkers:** Robert Curbeam, Christer Fuglesang
	Purpose: Second of four scheduled EVAs. Reconfigure electrical cables of power channels 2 and 3 and Z1 patch panel; relocate both CETA carts; install insulation on force moment sensors of ISS robotic arm; install four bags containing fluid line and pump module servicing equipment

Curbeam and Fuglesang performed the second EVA, which was the first of two intended to rewire the Station. Before they swapped the cable connections and began the rewiring process, the pair was forced to shut down certain systems such as communication gear, lights, ventilation fans, and backup computers. Then, while allowing the remainder of ISS systems to run on power generated from solar

11. "Mission Archives: STS-116," Jeanne Ryba, editor, accessed 22 March 2014, (*http://www.nasa.gov/mission_pages/shuttle/shuttlemissions/archives/sts-116.html*); "STS-116 Delivers Permanent Power," Jeanne Ryba, editor, accessed 22 March 2014, (*http://www.nasa.gov/mission_pages/shuttle/shuttlemissions/sts116/launch/sts116_summary.html*).

arrays installed during the STS-115 mission, Curbeam and Fuglesang rewired channels 2 and 3. The rewiring took less than 3 hours. By the end of this EVA, the ISS temporarily ran on power generated solely from the P4 truss. Furthermore, they already had one of the two external thermal control loops diverting excess heat into space, and DC-to-DC converter units regulating voltages throughout ISS. By taking far less time than the allotted 6 hours, Fuglesang and Curbeam were also able to reposition the two CETA carts on the S0 truss, mount tool bags for future spacewalks, and place a thermal cover over the Station's robotic arm.[12]

16 DECEMBER

2006 EVA 13	**Duration:** 7:31
World EVA 259	**Spacecraft/mission:** STS-116
U.S. EVA 162	**Crew:** Mark Polansky, William Oefelein, Sunita Williams, Joan Higginbotham, Nicholas Patrick, Robert Curbeam, Thomas Reiter (European Space Agency), Christer Fuglesang (European Space Agency)
Shuttle EVA 97	**Spacewalkers:** Robert Curbeam, Sunita Williams
	Purpose: Third of four scheduled EVAs. Reconfigure electrical cables of power channels 1 and 4, Z1 patch panel and Russian power; transfer service module MMOD protection panels to PMA-3; install grapple fixture on spare flex hose rotary coupler; manually aid retraction of P6 port solar array (completion deferred to next EVA)

To complete the rewiring, Curbeam accompanied Williams during the spacewalk. They reconfigured channels 1 and 4, which made possible future additions of European or Japanese laboratory modules. The astronauts also installed a grapple fixture on the Station's robotic arm and positioned bundles of Russian debris shields to be affixed to Zvezda. Before returning to Discovery, the pair also attempted to fix the P6 solar array, which the ISS crew had been unable to retract since the start of the STS-116 mission. The Station crew suspected that the array was caught on guide wires. Curbeam and Williams inspected the solar array and tried to clear any obstructing guide wires or grommets, and they both shook the array while the Station crew attempted to reel it in remotely, one bay at a time. Their combined efforts allowed the array to reach 65 percent retraction.[13]

---

12. "Mission Archives: STS-116," Jeanne Ryba, editor, accessed 22 March 2014, (*http://www.nasa.gov/mission_pages/shuttle/shuttlemissions/archives/sts-116.html*); "STS-116 Delivers Permanent Power," Jeanne Ryba, editor, accessed 22 March 2014, (*http://www.nasa.gov/mission_pages/shuttle/shuttlemissions/sts116/launch/sts116_summary.html*).

13. "Mission Archives: STS-116," Jeanne Ryba, editor, accessed 22 March 2014, (*http://www.nasa.gov/mission_pages/shuttle/shuttlemissions/archives/sts-116.html*); "STS-116 Delivers Permanent Power," Jeanne Ryba, editor, accessed 22 March 2014, (*http://www.nasa.gov/mission_pages/shuttle/shuttlemissions/sts116/launch/sts116_summary.html*).

FIGURE 7. **ISS 100th Construction Spacewalk.** Expedition 16 Commander Peggy Whitson works to re-connect cables on the 1A Beta Gimbal Assembly (BGA) during EVA 13. (NASA ISS016-E-17499)

18 DECEMBER

2006 EVA 14	**Duration:** 6:38
World EVA 260	**Spacecraft/mission:** STS-116
ESA EVA 8/U.S. EVA 163	**Crew:** Mark Polansky, William Oefelein, Sunita Williams, Joan Higginbotham, Nicholas Patrick, Robert Curbeam, Thomas Reiter (European Space Agency), Christer Fuglesang (European Space Agency)
Shuttle EVA 98	**Spacewalkers:** Robert Curbeam, Christer Fuglesang
	Purpose: Fourth of four scheduled EVAs. Manually retract P6 port solar array; inspect P6 starboard solar array; adjust new insulation of SSRMS force sensors

Curbeam and Fuglesang performed this impromptu EVA to repair for the final time the troublesome P6 solar array. Joan Higginbotham and Williams brought the two spacewalkers into position with the Space Station's robotic arm. Under William Oefelein's direction, they were eventually able to guide the array back into its blanket box completely. Lastly, Fuglesang and Curbeam secured the insulating thermal cover that they had placed on the ISS's robotic arm four days earlier. With Curbeam taking part in all four EVAs on the STS-116 mission, he set a new record for most spacewalks performed by one person during a single mission.[14]

22 December STS-116/Discovery landing

14. "Mission Archives: STS-116," Jeanne Ryba, editor, accessed 22 March 2014, (*http://www.nasa.gov/mission_pages/shuttle/shuttlemissions/archives/sts-116.html*); "STS-116 Delivers Permanent Power," Jeanne Ryba, editor, accessed 22 March 2014, (*http://www.nasa.gov/mission_pages/shuttle/shuttlemissions/sts116/launch/sts116_summary.html*).

2007

31 JANUARY

2007 EVA 1	**Duration:** 7:55
World EVA 261	**Spacecraft/mission:** International Space Station Expedition 14
U.S. EVA 164	**Crew:** Michael Lopez-Alegria (NASA), Mikhail Tyurin (Russian Space Agency), Thomas Reiter (European Space Agency), Sunita Williams (NASA)
Space Station EVA 133	**Spacewalkers:** Michael Lopez-Alegria, Sunita Williams
ISS EVA 28	**Purpose:** Establish permanent electrical and fluid connection between Destiny and Z1 truss; secure and cover P6 truss's radiator; remove one of two ammonia service jumpers; install Space Shuttle Power Transfer System

Spacewalkers Lopez-Alegria and Williams initially tackled separate tasks, and then collaborated on the final job. At an area near the Z1 base referred to as the "rats' nest," fluid lines ran between Destiny and the EAS. Because these cooling lines were deemed unnecessary, Lopez-Alegria was assigned to disconnect the cables and establish a permanent connection between the lab and the Z1 panel. He only had enough time to remove one of the two cables, which were later jettisoned in the summer. The retrieval of the remaining cable was postponed. He then installed the Space Shuttle Power Transfer System (SSPTS), a mechanism that conveyed power from the solar arrays to docked Space Shuttles. Meanwhile, Williams established a permanent electrical connection between Destiny and the Z1 truss. The pair then installed two winch bars and six cable cinches to secure the P6 starboard radiator and mounted a shroud over it. Upon reentering the Quest Airlock, the spacewalkers spent 25 minutes in sunlight in order to bake off any ammonia that may have gotten on their suits as a precaution to prevent ammonia contamination in the airlock.[1]

4 FEBRUARY

2007 EVA 2	**Duration:** 7:11
World EVA 262	**Spacecraft/mission:** International Space Station Expedition 14
U.S. EVA 165	**Crew:** Michael Lopez-Alegria (NASA), Mikhail Tyurin (Russian Space Agency), Thomas Reiter (European Space Agency), Sunita Williams (NASA)
Space Station EVA 134	**Spacewalkers:** Michael Lopez-Alegria, Sunita Williams
ISS EVA 29	**Purpose:** Establish an electrical and fluid connection between Destiny and the Z1 truss; retrieve the second ammonia service jumper; secure and stow the P6 truss's radiator; continue installation of SSPTS; photograph P6 solar array

---

1. "International Space Station: Station Crew Members Wind up Successful Spacewalk," Amiko Kauderer, editor, accessed 1 April 2014, (*http://www.nasa.gov/mission_pages/station/expeditions/expedition14/exp14_eva6.html*).

Astronauts Lopez-Alegria and Williams completed all scheduled tasks and an additional job planned for the following EVA. First, the pair returned to the "rats' nest" and retrieved the second of two cooling loops on Destiny. The first was removed during the previous spacewalk. They then installed the second set of two winch bars and six cable cinches. Lopez-Alegria and Williams mounted the first set during the previous EVA. In addition, by rewiring the Destiny lab, the astronauts finalized work with the EAS. With the permanent cooling system fully functioning, the early servicer only functioned as an emergency ammonia supply. Furthermore, the pair photographed the P6 solar wing, relocated spacewalk tools and equipment to the front of the Destiny lab, and continued rerouting the SSPTS cables. Time permitted the team to remove a thermal shade from a data transmitter called the multiplexer-demultiplexer.[2]

8 FEBRUARY

2007 EVA 3	**Duration:** 6:40
World EVA 263	**Spacecraft/mission:** International Space Station Expedition 14
U.S. EVA 166	**Crew:** Michael Lopez-Alegria (NASA), Mikhail Tyurin (Russian Space Agency), Thomas Reiter (European Space Agency), Sunita Williams (NASA)
Space Station EVA 135	**Spacewalkers:** Michael Lopez-Alegria, Sunita Williams
ISS EVA 30	**Purpose:** Remove and jettison two thermal shrouds on Rotary Joint Motor Controllers; remove and jettison two thermal shrouds from P6 truss; deploy Unpressurized Cargo Carrier Attachment System; remove launch locks off P5 truss; connect four cables between PMA-2 and SSPTS

During the last of three EVAs from the Quest Airlock, Lopez-Alegria and Williams completed all scheduled tasks and an additional assignment for a future EVA. They first used the CETA carts to transfer equipment to the P3 truss, and then detached two thermal shrouds from the Rotary Joint Motor Controllers and jettisoned them into space. They remained in the P6 truss worksite and detached two large thermal shades from bays 18 and 20. With the orientation of the Station, too much heat would be trapped under the shrouds. The spacewalkers compressed both shades into the size of an outdoor garbage can and jettisoned them. Next, Lopez-Alegria deployed an Unpressurized Cargo Carrier Attachment System (UCCAS) on the P3 truss, to prepare for the scheduled attachment of a cargo carrier. Meanwhile, Williams prepared for the attachment of the P6 truss by detaching the two remaining launch locks off the P5 truss. The spacewalkers then collaborated on the final task and connected four cables between the PMA-2 and the SSPTS. The latter transferred energy from the Station to docked Shuttles, allowing the visiting vehicles to stay longer. Additionally, Lopez-Alegria was able to photograph communication cables between the Station and the Shuttle. This EVA finalized the record-breaking use of the Quest Airlock. The crew exited the hatch for three spacewalks, the most by any crew thus far. This was Lopez-Alegria's 10th EVA, making him the record holder for most

2. "International Space Station: Crew Completes Scheduled Spacewalk Tasks, and More," Amiko Kauderer, editor, accessed 1 April 2014, (*http://www.nasa.gov/mission_pages/station/expeditions/expedition14/exp14_eva7.html*).

EVAs by an American. Furthermore, with four excursions, Williams became the most experienced female spacewalker.[3]

22 FEBRUARY

2007 EVA 4	**Duration:** 6:18
World EVA 264	**Spacecraft/mission:** International Space Station Expedition 14
Russian EVA 120/ U.S. EVA 167	**Crew:** Michael Lopez-Alegria (NASA), Mikhail Tyurin (Russian Space Agency), Thomas Reiter (European Space Agency), Sunita Williams (NASA)
Space Station EVA 136	**Spacewalkers:** Mikhail Tyurin, Michael Lopez-Alegria
ISS EVA 131	**Purpose:** Retract faulty Russian satellite navigational antenna; adjust ATV antenna; replace German experiment materials; photograph ATV reflectors and numerous experiments; attach two foot restraints outside Pirs airlock

Early in the final EVA, a faulty cooling system in Tyurin's suit caused his visor to fog up. The spacewalkers, however, continued investigating a malfunctioning navigational antenna on a Progress vehicle. The antenna, designed to guide the cargo carrier during docking, refused to withdraw during the 26 October 2006, docking. Tyurin and Lopez-Alegria unsuccessfully attempted to release the jammed antenna with a punch and hammer. They then severed the antenna's supportive struts and partially retracted the antenna. Wire ties secured the protruding parts. Also, the pair assessed and photographed an ATV antenna, a German experiment, various hardware connectors, and retention devices. They then bolted hinges on the Strela crane and, finally, mounted two foot restraints on the ladder by Pirs.[4]

7 April Soyuz-TMA 10/ISS Expedition 15 launch

21 April Soyuz-TMA 9/ISS Expedition 14 landing

30 MAY

2007 EVA 5	**Duration:** 5:25
World EVA 265	**Spacecraft/mission:** International Space Station Expedition 15
Russian EVA 121	**Crew:** Fyodor Yurchikhin, Oleg Kotov, Clayton Anderson (NASA), Sunita Williams (NASA)
Space Station EVA 137	**Spacewalkers:** Fyodor Yurchikhin, Oleg Kotov
ISS EVA 32	**Purpose:** Transfer MMOD panels from PMA-3 to SM; install five panels and ATV GPS antenna cable

3. "International Space Station: Spacewalkers Successfully Wrap Up Record Series," Amiko Kauderer, editor, accessed 1 April 2014, (*http://www.nasa.gov/mission_pages/station/expeditions/expedition14/exp14_eva8.html*).

4. "International Space Station: Spacewalkers Successfully Retract Progress Antenna," Amiko Kauderer, editor, accessed 10 April 2014, (*http://www.nasa.gov/mission_pages/station/expeditions/expedition14/exp14_eva17.html*).

Fyodor Yurchikhin and Oleg Kotov embarked on their first of two spacewalks aimed at armoring Zvezda with Service Module Debris Protection (SMDP). The cosmonauts exited the station through the Pirs airlock and split up. Kotov went to PMA-3 and Yurchikhin headed to the Strela 2 on Pirs. Kotov guided Yurchikhin to maneuver the crane to the SMDP stowage rack located on PMA-3. The rack contained 17 SMDP panels in total and was nicknamed the "Christmas Tree." Kotov attached the Christmas Tree to Strela 2 and Yurchikhin moved the Tree and Kotov back to a grapple fixture on Zvezda. Yurchikhin then joined Kotov on Zvezda's large diameter to reroute a GPS cable to be used with the European ATV scheduled for launch later in the year. The duo turned their attention back to the Christmas Tree and opened one bundle of five debris panels, which they affixed to the conical section between Zvezda's large and small diameters.[5]

6 JUNE

2007 EVA 6	**Duration:** 5:37
World EVA 266	**Spacecraft/mission:** International Space Station Expedition 15
Russian EVA 122	**Crew:** Fyodor Yurchikhin, Oleg Kotov, Clayton Anderson (NASA), Sunita Williams (NASA)
Space Station EVA 138	**Spacewalkers:** Fyodor Yurchikhin, Oleg Kotov
ISS EVA 33	**Purpose:** Install SM-FGB cable for local area network (LAN) and Biorisk science experiment; install 12 remaining MMOD shields on SM

Kotov and Yurchikhin installed a Russian scientific experiment on the outside of Pirs. Known as Biorisk, it analyzed how microorganisms affect structural materials in space. They then ran a reel of Ethernet cable along Zarya. It was the first of two strands of cable to be installed, and together, they were intended to increase the station's computing power. The two cosmonauts then returned to the Christmas Tree to gather the remaining 12 SMDP panels and secured them around Zvezda where they had attached five panels on 30 May.[6]

8 June STS-117/Atlantis launch

5. "Spacewalk Complete, Debris Panels Installed," Amiko Kauderer, editor, accessed 27 March 2014, (*http://www.nasa.gov/mission_pages/station/expeditions/expedition15/exp15_eva18.html*).

6. "Cosmonauts Wrap up Debris-Panel Spacewalk," Amiko Kauderer, editor, accessed 27 March 2014, (*http://www.nasa.gov/mission_pages/station/expeditions/expedition15/exp15_eva19.html*).

11 JUNE

2007 EVA 7	**Duration:** 6:15
World EVA 267	**Spacecraft/mission:** STS-117
U.S. EVA 168	**Crew:** Frederick Sturckow, Lee Archambault, Patrick Forrester, Danny Olivas, Clayton Anderson, James Reilly, Steven Swanson
Shuttle EVA 99	**Spacewalkers:** James Reilly, Danny Olivas
	Purpose: First of four scheduled EVAs. Connect electrical lines between Starboard Three (S3) and Starboard Four (S4) trusses to S1; release launch restraints

Prior to the start of the spacewalk, the STS-117 crew remotely transferred the 17.8-ton S3 and S4 trusses to the worksite. The heavy asymmetrical weight of the trusses caused the CMGs to go offline, and consequently the Station temporarily lost attitude. The crew's first EVA was delayed an hour by the expected loss. After reactivating the CMGs and increasing the attitude, James Reilly and Danny Olivas began the spacewalk. They focused on mounting the trusses and connecting electrical lines between the S1 truss and the two new trusses. The duo also removed six launch locks and prepared for the deployment of the S4 truss for the following excursion.[7]

13 JUNE

2007 EVA 8	**Duration:** 7:16
World EVA 268	**Spacecraft/mission:** STS-117
U.S. EVA 169	**Crew:** Frederick Sturckow, Lee Archambault, Patrick Forrester, Danny Olivas, Clayton Anderson, James Reilly, Steven Swanson
Shuttle EVA 100	**Spacewalkers:** Patrick Forrester, Steven Swanson
	Purpose: Second of four scheduled EVAs. Retract bays of P6 solar array to permit alpha joint rotation; remove launch locks; deploy braces and engage drive lock of S3 solar array alpha joint

With the unfurling of the solar array now attached to ISS's S3 and S4 trusses, Steven Swanson and Patrick Forrester were able to remove launch locks holding the array's Solar Alpha Rotary Joint in place. Swanson and Forrester then attempted to install a new drive-lock assembly into the Station. They could not carry out the installation successfully because flight controllers determined that commands they were trying to send to the drive-lock assembly were actually being received by an existing drive-lock assembly. Once flight controllers determined that the previously installed drive-lock assembly was still in a safe configuration, the two spacewalkers concluded their EVA by retracting one of the Station's older solar arrays to make room for the new one.[8]

---

7. "STS-117: 21st Space Station Flight," John F. Kennedy Space Center, accessed 29 March 2014, (*http://www.nasa.gov/pdf/177714main_STS-117.pdf*).

8. Ibid.

15 JUNE

2007 EVA 9	**Duration:** 7:58
World EVA 269	**Spacecraft/mission:** STS-117
U.S. EVA 170	**Crew:** Frederick Sturckow, Lee Archambault, Patrick Forrester, Danny Olivas, Clayton Anderson, James Reilly, Steven Swanson
Shuttle EVA 101	**Spacewalkers:** James Reilly, Danny Olivas
	Purpose: Third of four scheduled EVAs. Install vent valve on Lab for U.S. O_2 generator; secure loose thermal blanket on Shuttle's Orbital Maneuvering System (OMS); complete retraction of P6 solar array

Olivas was forced to begin the mission's third EVA by performing an impromptu repair on Atlantis's Orbital Maneuvering System pod. He reattached with staples and pins a 4 by 6 inch portion of its thermal blanket that peeled away during the Shuttle's launch. Reilly then installed an external hydrogen vent valve, delivered by Atlantis, onto Destiny. The vent is a component of the future oxygen generation system. Upon the completion of both astronauts' tasks, the duo helped to retract the Station's P6 solar array. Reilly and Olivas entered 28 commands in addition to the previously entered 17 to complete the retraction.[9]

17 JUNE

2007 EVA 10	**Duration:** 6:29
World EVA 270	**Spacecraft/mission:** STS-117
U.S. EVA 171	**Crew:** Frederick Sturckow, Lee Archambault, Patrick Forrester, Danny Olivas, Clayton Anderson, James Reilly, Steven Swanson
Shuttle EVA 102	**Spacewalkers:** Patrick Forrester, Steven Swanson
	Purpose: Fourth of four scheduled EVAs. Open H_2 vent for new O_2 generator; install video camera stand; verify drive lock configuration; remove additional restraints on the S3 solar array alpha joint; stow S3 drag link and keel pin; clear MT path; install LAN cable on Unity node

Swanson and Forrester retrieved a video camera from a stowage compartment located on Quest and installed it alongside the S3 truss. They then returned to the second drive-lock assembly that they installed four days earlier to ensure that it was indeed working under the correct configuration. Once verified, they removed the remainder of the launch locks holding down the SARJ. With the SARJ now able to rotate S3 freely, the spacewalkers cleared a path along S3 for the Mobile Base System (MBS). With the work moving ahead of schedule, Swanson and Forrester also performed the get-ahead task. They installed a LAN cable on the Unity node. They also opened the vent valve, which Reilly had installed on the mission's previous EVA, and attached two orbital debris shield panels on ISS's service

9. Ibid.

module. They were successful in the attachment of the shield panels, but they were forced to use tethered hooks in place of some of the original MMOD shield fasteners.[10]

22 June STS-117/Atlantis landing

23 JULY

2007 EVA 11	**Duration:** 7:41
World EVA 271	**Spacecraft/mission:** International Space Station Expedition 15
Russian EVA 123/ U.S. EVA 172	**Crew:** Fyodor Yurchikhin, Oleg Kotov, Clayton Anderson (NASA), Sunita Williams (NASA)
Space Station EVA 139	**Spacewalkers:** Clayton Anderson, Fyodor Yurchikhin
ISS EVA 34	**Purpose:** Install TV camera stanchion at S0-P1; reconfigure S-Band Antenna Assembly power supply; install foot restraint on Canadarm2; disconnect and jettison EAS and Flight Releasable Attachment Mechanism; clean CBM

Exiting the ISS from the Quest airlock, Clayton Anderson and Yurchikhin installed a TV camera stanchion on the S0-P1 truss. Anderson then reconfigured an S-Band Antenna Assembly power supply and mounted a foot restraint on the Canadarm2. Meanwhile, Yurchikhin replaced a circuit breaker for the main truss's Mobile Transporter rail car. Afterward, Yurchikhin and Anderson reconvened at the S0-P1 truss to remove some flight support equipment and a Flight Releasable Attachment Mechanism. Anderson, while attached to the newly installed foot restraint on Canadarm2, jettisoned them, and Yurchikhin moved to Z1 where he began to disconnect the EAS. Both spacewalkers proceeded to the P6 truss to undo EAS's remaining connections to the Station. Once finished, Anderson grabbed the EAS and Kotov maneuvered him and the arm to the bottom of the main truss to jettison the unit by pushing it in the opposite direction of the Station's orbital path. After the jettison, Yurchikhin and Anderson cleaned the CBM to prepare for the future relocation of the PMA-3 to that location, which itself was in preparation for the addition of the Harmony node. Ahead of schedule, the pair was also able to move an auxiliary equipment bag from P6 to Z1 and remove a GPS antenna and some bolts on two fluid trays on the S0 truss.[11]

8 August STS-118/Endeavour launch

10. Ibid.

11. "Station Crew Winds Up Ammonia Reservoir Jettison Spacewalk," Amiko Kauderer, editor, accessed 6 April 2014, (*http://www.nasa.gov/mission_pages/station/expeditions/expedition15/e15_eva_072307.html*).

11 AUGUST

2007 EVA 12	**Duration:** 6:17
World EVA 272	**Spacecraft/mission:** STS-118
Canadian EVA 4/ U.S. EVA 173	**Crew:** Scott Kelly, Charles Hobaugh, Barbara Morgan, Rick Mastracchio, Tracy Caldwell, Alvin Drew, Dave Williams (Canadian Space Agency)
Shuttle EVA 103	**Spacewalkers:** Rick Mastracchio, Dave Williams
	Purpose: First of four scheduled EVAs. Install Starboard Five (S5) truss and connect S4 and S5 electrical lines; relocate radiator grapple fixture on S5; remove S5 and S6 launch locks on S5 and open the S5 and S6 capture latch; secure the auto retracted P6 radiator

On STS-118's first EVA, the crew focused on work on the solar arrays. From within the Station, Charles Hobaugh and Clayton Anderson moved the S5 spacer into position to be attached to the ISS. Meanwhile, spacewalkers Rick Mastracchio and Dave Williams provided guidance from outside the ISS. Once in position, Mastracchio and Williams permanently attached S5 with its automatic latches and by securing its primary structural bolts to the Station. They then retracted the forward heat-rejecting radiator on P6 so that the truss could be moved during a later Shuttle mission. Having completed these tasks ahead of schedule, the spacewalkers removed the launch locks from S5 and S6 and opened both of their capture latches as well to make future jobs run more smoothly.[12]

13 AUGUST

2007 EVA 13	**Duration:** 6:28
World EVA 273	**Spacecraft/mission:** STS-118
Canadian EVA 5/ U.S. EVA 174	**Crew:** Scott Kelly, Charles Hobaugh, Barbara Morgan, Rick Mastracchio, Tracy Caldwell, Alvin Drew, Dave Williams (Canadian Space Agency)
Shuttle EVA 104	**Spacewalkers:** Rick Mastracchio, Dave Williams
	Purpose: Second of four scheduled EVAs. Replace failed CMG 3 and stow the failed unit on pallet outside ISS airlock for return on future Shuttle mission; photograph MISSE 3 and 4 experiments

With the aid of the crew operating Canadarm2 and Endeavour's robotic arm, Mastracchio and Williams replaced one of the Station's gyroscopes. The failed CMG had been out of operation since October of 2006. Mastracchio and Williams installed the 600-pound (272.16-kilogram) replacement unit on the Z1 segment of ISS's truss. They carried the failed unit to a stowage compartment on the exterior of the Space Station. It was to be returned to Earth on a later mission.[13]

---

12. "STATUS REPORT: STS-118-06," Amiko Nevills, editor, accessed 7 April 2014, (*http://www.nasa.gov/mission_pages/shuttle/shuttlemissions/sts118/news/STS-118-07.html*); "STS-118: 22nd Space Station Flight," John F. Kennedy Space Center, accessed 7 April 2014, (*http://www.nasa.gov/pdf/182309main_STS-118W.pdf*).

13. "STATUS REPORT: STS-118-11," Amiko Nevills, accessed 7 April 2014, (*http://www.nasa.gov/mission_pages/shuttle/shuttlemissions/sts118/news/STS-118-11.html*); "STS-118: 22nd Space Station Flight," John F. Kennedy Space Center, accessed 7 April 2014, (*http://www.nasa.gov/pdf/182309main_STS-118W.pdf*).

15 AUGUST

2007 EVA 14	**Duration:** 5:28
World EVA 274	**Spacecraft/mission:** STS-118
U.S. EVA 175	**Crew:** Scott Kelly, Charles Hobaugh, Barbara Morgan, Rick Mastracchio, Tracy Caldwell, Alvin Drew, Dave Williams (Canadian Space Agency)
Shuttle EVA 105	**Spacewalkers:** Rick Mastracchio, Clayton Anderson (ISS Expedition 15 crew member)
	Purpose: Third of four scheduled EVAs. Relocate the S-band antenna from P6 to P; install new signal processor and transponder on P1; retrieve P6 transponder; relocate CETA carts

STS-118's third EVA began at 9:38 a.m. with Anderson and Mastracchio moving the S-band antenna sub-assembly from P6 to P1. During the spacewalk, the duo also monitored ISS's robotic arm as Hobaugh and Kotov used it to move two CETA carts. Moving the carts would allow for relocation of the solar array segment to its permanent location during the STS-120 mission. At 1:45 p.m., Mastracchio noticed that his glove had a hole in its second of five layers. The hole caused no leakage and ultimately brought no harm to Mastracchio but he nonetheless entered the Quest Airlock as a precaution. Anderson continued working and was able to complete their task of installing a new transponder on P1 and retrieving P6's transponder. The spacewalkers were originally scheduled to retrieve MISSE 3 and 4, but Mastracchio's early return to the airlock forced them to postpone its retrieval to the mission's next EVA.[14]

18 AUGUST

2007 EVA 15	**Duration:** 5:02
World EVA 275	**Spacecraft/mission:** STS-118
Canadian EVA 6/ U.S. EVA 176	**Crew:** Scott Kelly, Charles Hobaugh, Barbara Morgan, Rick Mastracchio, Tracy Caldwell, Alvin Drew, Dave Williams (Canadian Space Agency)
Shuttle EVA 106	**Spacewalkers:** Dave Williams, Clayton Anderson (ISS Expedition 15 crew member)
	Purpose: Fourth of four scheduled EVAs. Install orbiter inspection boom restraints on S1; engage S-band antenna gimbal locks on Z1; retrieve MISSE 3 and 4; install antennas for external wireless sensors on lab

Anderson and Williams completed the final EVA of the mission by installing the external wireless instrumentation system antenna and attaching a stand on the Shuttle to hold up its robotic arm's extension boom. The two astronauts also were able to retrieve the two MISSE containers. Originally, they were scheduled to clean up and secure debris shields to ISS and move a toolbox to a more centralized

14. "STATUS REPORT: STS-118-11," Amiko Nevills, editor, accessed 7 April 2014, (*http://www.nasa.gov/mission_pages/shuttle/shuttlemissions/sts118/news/STS-118-15.html*); "STS-118: 22nd Space Station Flight," John F. Kennedy Space Center, accessed 7 April 2014, (*http://www.nasa.gov/pdf/182309main_STS-118W.pdf*).

location on the Station, but these tasks were delayed to a later mission to allow them to close the airlock hatch a day earlier.[15]

21 August STS-118/Endeavour landing

10 October Soyuz-TMA 11/ISS Expedition 16 launch

21 October Soyuz-TMA 10/ISS Expedition 15 landing

23 October STS-120/Discovery launch

26 OCTOBER

2007 EVA 16
World EVA 276
U.S. EVA 177
Shuttle EVA 107

Duration: 6:14

Spacecraft/mission: STS-120

Crew: Pamela Melroy, George Zamka, Scott Parazynski, Douglas Wheelock, Stephanie Wilson, Daniel Tani (began tour on ISS), Clayton Anderson (returned to Earth), Paolo Nespoli (European Space Agency)

Spacewalkers: Scott Parazynski, Douglas Wheelock

Purpose: First of four scheduled EVAs. Stow failed S-band antenna; release Node 2 Harmony launch power cable; fasten Harmony grapple fixture; secure Harmony window cover that opened during launch; detach connections between Z1 and P6 trusses

Discovery's docking on the ISS marked the first engagement between two spacecraft commanded by women: Pamela Melroy and Peggy Whitson. During the crew's first of four missions, Parazynski and Wheelock retrieved a failed Station antenna and stowed it in Discovery. They also prepared for the relocation of the P6 truss by disconnecting fluid lines between the P6 and the Z1 trusses. The astronauts mainly focused on preparations to install the 31,500-pound (14,288.4-kilogram) Harmony node, which is also known as "Node 2." The pair released Harmony's launch power cables and temporarily secured a grapple fixture, as well as the window cover that unexpectedly opened during launch. Parazynski and Wheelock then attached the node to its temporary location until more preparations were later completed. Though it was built in Turin, Italy, by Thales Alenia Space, Harmony became the United States's fourth ISS module when it joined Destiny, Quest, and Unity. The new addition acted as a pressurized module, connecting cargo spacecrafts and the laboratories. Harmony circulates energy from the trusses to the United States's Destiny lab, ESA's Columbus lab, and Japan's Kibo lab. Furthermore, Harmony's exterior functions as a work platform for Canadarm2.[16]

15. "STATUS REPORT: STS-118-21," Amiko Nevills, editor, accessed 7 April 2014, (*http://www.nasa.gov/mission_pages/shuttle/shuttlemissions/sts118/news/STS-118-21.html*); "STS-118: 22nd Space Station Flight," John F. Kennedy Space Center, accessed 7 April 2014, (*http://www.nasa.gov/pdf/182309main_STS-118W.pdf*).

16. "Harmony," Jerry Wright, editor, accessed 20 April 2014, (*http://www.nasa.gov/mission_pages/station/structure/elements/node2.html#*.U7Nc5bHvaJs); "STS-120 (23rd Space Station Flight): *Discovery*," John F. Kennedy Space Center, accessed 20 April 2014, (*http://www.nasa.gov/pdf/216375main_STS-120.pdf*).

28 OCTOBER

2007 EVA 17
World EVA 277
U.S. EVA 178
Shuttle EVA 108

Duration: 6:33

Spacecraft/mission: STS-120

Crew: Pamela Melroy, George Zamka, Scott Parazynski, Douglas Wheelock, Stephanie Wilson, Daniel Tani, Clayton Anderson (returned to Earth), Paolo Nespoli (European Space Agency)

Spacewalkers: Scott Parazynski, Daniel Tani

Purpose: Second of four scheduled EVAs. Detach P6 umbilical cords and bolts; install Harmony EVA aids; remove Harmony berthing pins; inspect Solar Array Alpha Joint and collect metallic debris sample; configure S1 radiator release

Parazynski and Daniel Tani further prepared for the relocation of the P6 truss by disconnecting umbilical cables and attachments bolts to the Z1 truss. Tani then examined the SARJ, a device that rotates the solar arrays. The joint experienced increased vibrations and power usage. The spacewalkers discovered metallic debris under its insulation covers. Use of the SARJ was limited until samples of the debris were further studied. Parazynski and Tani then finalized the installation of Harmony. They installed the module's grappling fixture, handrails, and foot restraint sockets. The pair also removed Harmony's berthing mechanism pins, though they lost one. They finalized the spacewalk with the reconfiguration of the S1 truss connectors, allowing the truss's radiator to be controlled from the ground in the future.[17]

30 OCTOBER

2007 EVA 18
World EVA 278
U.S. EVA 179
Shuttle EVA 109

Duration: 7:08

Spacecraft/mission: STS-120

Crew: Pamela Melroy, George Zamka, Scott Parazynski, Douglas Wheelock, Stephanie Wilson, Daniel Tani, Clayton Anderson (returned to Earth), Paolo Nespoli (European Space Agency)

Spacewalkers: Scott Parazynski, Douglas Wheelock

Purpose: Third of four scheduled EVAs. Inspect Solar Array Alpha Joint; install P6 truss

Parazynski and Wheelock finalized the installation of the P6 truss by connecting four cables and four bolts between P6 and P5. They also removed the P6 truss's thermal blanket and its radiator restraints. Parazynski then examined the SARJ and did not identify any new metallic debris by the vents. At the conclusion of the EVA, the astronauts attempted to deploy the new truss's solar arrays. However, due to a tear, it only deployed 80 percent. The opening and use of the solar arrays were ceased until

17. "Space Shuttle Mission Summary," Robert D. Legler, Floyd V. Bennett," accessed 20 April 2014, (*http://www.jsc.nasa.gov/history/reference/TM-2011-216142.pdf*); "STS-120 (23rd Space Station Flight): *Discovery*," John F. Kennedy Space Center, accessed 20 April 2014, (*http://www.nasa.gov/pdf/216375main_STS-120.pdf*).

the tear was investigated further. Upon reentering the airlock, Wheelock identified a puncture in his right glove. After the EVA, the glove was retired due to thermal micrometeoroid garment damage. The astronauts also later realized they forgot a camera outside.[18]

3 NOVEMBER

2007 EVA 19	**Duration:** 7:19
World EVA 279	**Spacecraft/mission:** STS-120
U.S. EVA 180 **Shuttle EVA 110**	**Crew:** Pamela Melroy, George Zamka, Scott Parazynski, Douglas Wheelock, Stephanie Wilson, Daniel Tani, Clayton Anderson (returned to Earth), Paolo Nespoli (European Space Agency)
	Spacewalkers: Scott Parazynski, Douglas Wheelock
	Purpose: Fourth of four scheduled EVAs. Inspect and repair damaged P6 solar array; install five reinforcements to P6 truss; retrieve two foot restraints with sharp edges; retrieve camera left out during previous EVA

The crew's last spacewalk focused on repairing a tear in the newly installed P6 solar array. Parazynski and Wheelock rode the Canadarm2 to the array and examined the tear. Parazynski cut frayed hinges and wires. He then installed five stabilizers, which were made in the Station and resembled cufflinks, to strengthen the stability of the damaged panel hinges. The pair also retrieved the two foot restraints with sharp edges, which they suspected caused the spacesuit glove damage. The repairs were approved and ground controllers completed the successful deployment of the solar array.[19]

7 November STS-120/Discovery landing

9 NOVEMBER

2007 EVA 20	**Duration:** 6:55
World EVA 280	**Spacecraft/mission:** International Space Station Expedition 16
Russian EVA 124/ U.S. EVA 181 **Space Station EVA 140**	**Crew:** Peggy Whitson (NASA), Clayton Anderson (NASA), Leopold Eyharts (European Space Agency), Yuri Malenchenko (Russian Space Agency), Garrett Reisman (NASA), Daniel Tani (NASA)
ISS EVA 35	**Spacewalkers:** Peggy Whitson, Yuri Malenchenko
	Purpose: Prepare PMA-2 for relocation to Node 2; disconnect electrical cables; remove light on lab and thermal cover from Node 2 forward port; connect power/data cables for Node 2 grapple fixture; connect power cable for service module power; replace failed RPCM; reconfigure FGB-PMA1 jumpers; retrieve S-band signal processor; relocate ammonia vent tool bag on airlock

18. Ibid.

19. Ibid.

Wearing American spacesuits, astronaut Whitson and cosmonaut Malenchenko departed the ISS from the Quest Airlock to begin the first EVA of Expedition 16. Whitson was the mission's lead spacewalker, and they started by disconnecting the Station to Shuttle Power Transfer System cables as well as eight other cables between Destiny and PMA-2. To also allow for the Harmony Node move, Whitson removed a light on Destiny and took it back to the airlock and both spacewalkers undid connectors on rigid umbilicals on Destiny's exterior and stowed them on and around the lab. The light would later be reinstalled along with the connectors following the installation of fluid umbilical trays on the lab. Malenchenko concluded Destiny's preparation for the Node's relocation by installing caps on receptacles left open from the disconnection of the PMA cables. They then moved to Harmony, and Whitson finished making connections for a power and data grapple fixture that would provide a base on Harmony for Canadarm2. Meanwhile, Malenchenko removed and then replaced a failed RPCM. Then, to prepare for the installation of PMA-2, the duo moved to the outboard end of the Node to remove the CBM cover. Upon removing the cover, they bundled it and secured it with wire ties for later disposal in a Progress cargo carrier. Malenchenko and Whitson then split apart at Z1, where Malenchenko reconfigured its power system by removing an electrical jumper on its rear, and Whitson worked on another reconfiguration down at the Z1's base, or the "rat's nest." They finished off the EVA with Whitson bringing a base-band signal processor box to the Quest Airlock and Malenchenko transferring tools between two bags and leaving one on S0 to use in future spacewalks.[20]

20 NOVEMBER

2007 EVA 21
World EVA 281
U.S. EVA 182
Space Station EVA 141
ISS EVA 36

Duration: 7:16

Spacecraft/mission: International Space Station Expedition 16

Crew: Peggy Whitson (NASA), Clayton Anderson (NASA), Leopold Eyharts (European Space Agency), Yuri Malenchenko (Russian Space Agency), Garrett Reisman (NASA), Daniel Tani (NASA)

Spacewalkers: Peggy Whitson, Daniel Tani

Purpose: Install starboard thermal cooling lines on Node 2; connect electrical cables to PMA-2

Whitson and Tani set out on this EVA to begin moving the Harmony Node, which Whitson and Malenchenko did preparation work for in the spacewalk 11 days earlier. Whitson began by returning to the exterior of Destiny and removing venting and stowing an ammonia jumper near where she had previously removed a light. Tani retrieved the tool bag left by Malenchenko on the previous EVA and removed fluid caps on Destiny. Whitson and Tani's removal of different parts allowed for the installation of a permanent ammonia-cooling loop on a fluid tray on the lab's exterior. Tani then reconfigured a circuit on a squib-firing unit that was used to deploy a radiator on P1 on 15 November. The duo then met at S0 to begin the move of the loop A fluid tray from there to the Harmony Node. The tray is 300 pounds (136.08 kilograms) and 18.5 feet (5.64 meters) long, and the two astronauts used a relay technique in moving it. One would move ahead, attaching tethers to make the tray ready for

20. "Station Spacewalk Prepares for PMA, Harmony Moves," Amiko Kauderer, editor, accessed 21 April 2014, (*http://www.nasa.gov/mission_pages/station/expeditions/expedition16/eva_pma_harmony.html*).

receiving, and the other would move farther ahead to be ready for the next handoff. After delivering the tray to Harmony, they bolted it down and connected its fluid lines, with two connecting to S0, two to the tray itself, and two in between them. Tani then mated 11 avionics lines on the port side of the Node. Whitson, meanwhile, configured the heater cables on Harmony and then performed the last scheduled task of the spacewalk, which was mating electrical umbilicals by hooking up four electrical harnesses linking PMA-2 at the end of Harmony to power from ISS. Tani then completed a get-ahead task of connecting five starboard avionics umbilicals to Harmony. Tani and Whitson also connected redundant umbilicals to PMA-2. They also began connecting cables at PMA-2 to transfer power from the Shuttle to the Station. Once the spacewalkers entered the airlock, they underwent precautionary decontamination procedures due to working on the ammonia lines.[21]

24 NOVEMBER

2007 EVA 22
World EVA 282
U.S. EVA 183
Space Station EVA 142
ISS EVA 37

Duration: 7:04

Spacecraft/mission: International Space Station Expedition 16

Crew: Peggy Whitson (NASA), Clayton Anderson (NASA), Leopold Eyharts (European Space Agency), Yuri Malenchenko (Russian Space Agency), Garrett Reisman (NASA), Daniel Tani (NASA)

Spacewalkers: Peggy Whitson, Daniel Tani

Purpose: Install port thermal cooling lines on Node 2; connect electrical cables to PMA-2; inspect and photograph starboard Solar Array Rotary Joint (SARJ)

Whitson and Tani reunited to venture outside the Station and continue the work they started four days earlier. Again, they removed a temporary ammonia jumper and fluid caps, this time to allow for the transfer of the loop B fluid tray from S0 to the port avionics tray atop of Destiny. They again relayed the tray to each other to deliver to the installation point. Once there, they bolted it down and connected its six fluid lines and its two heater cables. Whitson then moved to the starboard side of Harmony to do work that would prepare for its attachment to the ESA laboratory, Columbus, during the STS-122 mission in the December 2007. She removed a thermal cover over the centerline berthing camera mechanism and eight launch restraints over latch petals from a common berthing mechanism. She then finished the SSPTS cable connection that she and Tani had started making on the previous EVA. Tani, meanwhile, removed one of the covers on the starboard SARJ on the right side of the Station's main truss. He took photographs and collected metallic shavings from the surface. Whitson then joined him at the starboard SARJ for analysis. They observed that some surfaces under the cover were abraded and that the joint showed unexpected vibration and increased power consumption. The photos and shavings would be brought back for further inspection, and they stowed the SARJ cover inside the airlock so they could observe the joint in rotation with Canadarm2's video camera. Whitson then reinstalled the light on Destiny, which she had removed on the spacewalk of November 9. The

21. "Spacewalkers Harmonize on Node Hookup Tasks," John Ira Petty, accessed 21 April 2014, (*http://www.nasa.gov/mission_pages/station/expeditions/expedition16/exp16_eva_112007.html*).

EVA concluded with Tani relocating a tool bag and an articulating portable foot restraint in preparation for future work on ISS.[22]

18 DECEMBER

2007 EVA 23	**Duration:** 6:56
World EVA 283	**Spacecraft/mission:** International Space Station Expedition 16
U.S. EVA 184	**Crew:** Peggy Whitson (NASA), Clayton Anderson (NASA), Leopold Eyharts (European Space Agency), Yuri Malenchenko (Russian Space Agency), Garrett Reisman (NASA), Daniel Tani (NASA)
Space Station EVA 143	**Spacewalkers:** Daniel Tani, Peggy Whitson
ISS EVA 38	**Purpose:** Inspect starboard solar array alpha and beta joints; retrieve one trundle bearing from alpha joint for in-cabin and ground assessments

This was the 100th spacewalk devoted to construction and maintenance of the ISS. Whitson also set a record for the most cumulative spacewalk time by a woman during this EVA. By its end, she had amassed a total of 32 hours and 36 minutes of spacewalk time. Tani and Whitson began this EVA with the inspection of Beta Gimbal Assembly 1A (BGA 1A) because its primary power was lost when three of its circuit breakers tripped. Upon initial inspection, they did not find any obvious damage, but they disconnected two cables to facilitate ground tests. With the cables disconnected, the circuits were still observed to be closed, ruling out the possibility that the cables were responsible for the power failure. Whitson later reconnected the cables. The astronauts then moved to the SARJ to remove two drive lock assembly covers and inspect race rings and bearings underneath them. They also looked underneath most of the 22 covers over SARJ, to make inspections similar to the one that Tani had done during the 24 November spacewalk. They found debris, dust, and contamination under most of the covers, just as they found before. They took more photographs and collected more debris samples to bring back to the Station for further analysis. Lastly, before returning to the airlock, they removed Trundle Bearing Assembly #5 to bring with them for assessment inside the Station.[23]

---

22. "Spacewalkers Complete More Harmony Hookup Work," John Ira Petty, editor, accessed 21 April 2014, (*http://www.nasa.gov/mission_pages/station/expeditions/expedition16/exp16_eva_112407.html*).

23. "Spacewalkers Find No Solar Wing Smoking Gun," John Ira Petty, editor, accessed 21 April 2014, (*http://www.nasa.gov/mission_pages/station/expeditions/expedition16/exp16_eva_121807.html*).

2008

30 JANUARY

2008 EVA 1	**Duration:** 7:10
World EVA 284	**Spacecraft/mission:** International Space Station Expedition 16
U.S. EVA 185	**Crew:** Peggy Whitson (NASA), Clayton Anderson (NASA), Leopold Eyharts (European Space Agency), Yuri Malenchenko (Russian Space Agency), Garrett Reisman (NASA), Daniel Tani (NASA)
Space Station EVA 144	**Spacewalkers:** Peggy Whitson, Daniel Tani
ISS EVA 39	**Purpose:** Replace the Bearing Motor Roll Ring Module (BMRRM) and inspect SARJ

This was the last of five EVAs that took place during Expedition 16. Whitson and Tani replaced the BMRRM, also known as the "broom," which is the motor that rotates the Station's solar wings. The duo retrieved the new BMRRM from stowage in PMA-3. Once the astronauts installed the new motor, they observed as flight controllers rotated the solar wings a full 360 degrees. With the wings appearing to be working in perfect order with the replacement BMRRM, the spacewalkers returned to the starboard SARJ on the main truss. Tani and Whitson this time removed eight more covers on the joint to continue their inspection for contamination and debris.[1]

7 February STS-122/Atlantis launch

11 FEBRUARY

2008 EVA 2	**Duration:** 7:58
World EVA 285	**Spacecraft/mission:** STS-122
U.S. EVA 186	**Crew:** Stephen Frick, Alan Poindexter, Rex Walheim, Stanley Love, Leland Melvin, Hans Schlegel (European Space Agency), Leopold Eyharts (ESA, began tour on ISS)
Shuttle EVA 111	**Spacewalkers:** Rex Walheim, Stanley Love
	Purpose: First of three scheduled EVAs. Begin installation of Columbus Module; complete preparatory work to replace nitrogen tank

The ISS crew and the Shuttle crew worked together to install the Columbus laboratory, the ESA's largest contribution to the ISS. The laboratory may contain up to 10 payload racks—eight within the sidewalls and two in the ceiling. The racks function as individual laboratories. Each is equipped with a

1. "Spacewalkers Replace Solar Wing Motor, John Ira Petty, editor, accessed 21 April 2014, (*http://www.nasa.gov/mission_pages/station/expeditions/expedition16/exp16_eva_013008.html*).

cooling system, as well as data and video connections for researchers on Earth. Columbus was designed to be minimalistic with the utmost efficiency—maximum research with minimal space. While the laboratory was still docked in Atlantis's payload bay, Stanley Love and Rex Walheim installed a grapple fixture, removed window and berthing covers, and disconnected heating cables between Columbus lab and the Shuttle. They then readied its data and power lines to be mated with the ISS. Then, Leland Melvin, Leopold Eyharts, and Tani remotely transported Columbus to the Station's starboard side, via the Canadarm2. The spacewalkers concluded the EVA with the loosening of the Station's nitrogen tank, which was scheduled for replacement.[2]

13 FEBRUARY

2008 EVA 3

World EVA 286

ESA EVA 9/U.S. EVA 187

Shuttle EVA 112

Duration: 6:45

Spacecraft/mission: STS-122

Crew: Stephen Frick, Alan Poindexter, Rex Walheim, Stanley Love, Leland Melvin, Hans Schlegel (European Space Agency), Leopold Eyharts (ESA, began tour aboard the ISS)

Spacewalkers: Rex Walheim, Hans Schlegel

Purpose: Second of three scheduled EVAs. Replace nitrogen tank on P1 truss; install trunnion covers on Columbus; repair Destiny debris shield

Hans Schlegel and Walheim completed the work from the previous spacewalk and replaced the Station's nitrogen tank on the P1 truss. Using the Canadarm2, they stored the old tank into Atlantis's payload bay for return to Earth. Additionally, the spacewalkers installed trunnion covers on Columbus and made small repairs on the Destiny debris shield.[3]

15 FEBRUARY

2008 EVA 4

World EVA 287

U.S. EVA 188

Shuttle EVA 113

Duration: 7:25

Spacecraft/mission: STS-122

Crew: Stephen Frick, Alan Poindexter, Rex Walheim, Stanley Love, Leland Melvin, Hans Schlegel (European Space Agency), Leopold Eyharts (ESA, began tour aboard the ISS)

Spacewalkers: Rex Walheim, Stanley Love

Purpose: Third of three scheduled EVAs. Install two ESA experiments; mount nine handrails, two foot restraints, and a keep pin cover on Columbus; store failed CMG in Atlantis; photograph sharp edge on Quest Airlock handrail

---

2. "*Columbus* Laboratory," European Space Agency, accessed 1 June 2014, (*http://www.nasa.gov/pdf/216164main_STS-122.pdf*); "STS-122 (24th Space Station Flight): *Atlantis*," John F. Kennedy Space Center, accessed 1 June 2014, (*http://www.esa.int/Our_Activities/Human_Spaceflight/Columbus/Columbus_laboratory*).

3. "STS-122 (24th Space Station Flight): *Atlantis*," John F. Kennedy Space Center, accessed 1 June 2014, (*http://www.esa.int/Our_Activities/Human_Spaceflight/Columbus/Columbus_laboratory*).

Using the Canadarm2, Stanley Love and Rex Walheim installed two ESA experiments, Sun Monitoring on the External Payload Facility of Columbus (SOLAR) and European Technology Exposure Facility (EuTEF), to Columbus's exterior. They also mounted nine handrails, two foot restraint sockets, and a keel pin cover on the laboratory. The pair then retrieved and stored a failed CMG in Atlantis, for return to Earth. Love and Walheim also examined and photographed a handrail on the Quest Airlock's handrail. A sharp edge on the handrail was suspected in an accidental cut to the gloves.[4]

20 February STS-122/Atlantis landing

11 March STS-123/Endeavour launch

13 MARCH

2008 EVA 5	**Duration:** 7:01
World EVA 288	**Spacecraft/mission:** STS-123
U.S. EVA 189	**Crew:** Dominic Gorie, Gregory Johnson, Garrett Reisman (began tour aboard the ISS), Robert Behnken, Michael Foreman, Richard Linnehan, Takao Doi (JAXA), Leopold Eyharts (ESA, returned to Earth)
Shuttle EVA 114	**Spacewalkers:** Richard Linnehan, Garrett Reisman
	Purpose: First of five scheduled EVAs. Install Kibo Laboratory; install tools on Canadarm2; mount Centerline Berthing Camera System on Harmony

Linnehan, an experienced spacewalker, led the first of five EVAs. With Garrett Reisman, he first installed and powered the Japanese laboratory Kibo, which translates as "Hope." The laboratory is Japan's first spacecraft facility. It houses experiments relating to biology, medicine, communications, biotechnology, and observations of Earth. Furthermore, Kibo has a scientific airlock, allowing experiments to be transferred to the exterior and exposed to the space environment. To install Kibo, first-time spacewalker Garrett Reisman and Linnehan disconnected heater cables between the laboratory and Endeavour, then detached its berthing mechanism covers. From within the Station, Doi and Gorie operated the Canadarm2 and transferred Kibo to its temporary location on Harmony. The spacewalkers then focused on mounting the Centerline Berthing Camera System on Harmony. The camera offers live images to assist in the installation of new modules.[5]

---

4. "STS-122 (24th Space Station Flight): *Atlantis*," John F. Kennedy Space Center, accessed 1 June 2014, (*http://www.esa.int/Our_Activities/Human_Spaceflight/Columbus/Columbus_laboratory*).

5. "All Aboard: The Station Goes Global," National Aeronautics and Space Administration, accessed 5 June 2014, (*http://www.nasa.gov/pdf/215905main_sts123_press_kit_b.pdf*); "International Space Station: Kibo Laboratory," Jerry Wright, editor, accessed 5 June 2014, (*http://www.nasa.gov/mission_pages/station/structure/elements/jem.html#.Vuw04-cuJ2B*).

15 MARCH

2008 EVA 6	**Duration:** 7:09
World EVA 289	**Spacecraft/mission:** STS-123
U.S. EVA 190	**Crew:** Dominic Gorie, Gregory Johnson, Garrett Reisman (began tour aboard the ISS), Robert Behnken, Michael Foreman, Richard Linnehan, Takao Doi (JAXA), Leopold Eyharts (ESA, returned to Earth)
Shuttle EVA 115	**Spacewalkers:** Richard Linnehan, Michael Foreman
	Purpose: Second of five scheduled EVAs. Assemble Dextre arms

Linnehan performed the second of five EVAs with first-time spacewalker Michael Foreman. The pair installed the Canadian-built Special Purpose Dexterous Manipulator (SPDM), which is more commonly known as Dextre, a sophisticated robot that performs small maintenance work and repairs. Its assistance will make EVAs more efficient by allowing spacewalkers to spend more time on experiments, rather than maintenance routines. The spacewalkers removed covers and assembled both arms on Dextre.[6]

17 MARCH

2008 EVA 7	**Duration:** 6:35
World EVA 290	**Spacecraft/mission:** STS-123
U.S. EVA 191	**Crew:** Dominic Gorie, Gregory Johnson, Garrett Reisman (began tour aboard the ISS), Robert Behnken, Michael Foreman, Richard Linnehan, Takao Doi (JAXA), Leopold Eyharts (ESA, returned to Earth)
Shuttle EVA 116	**Spacewalkers:** Richard Linnehan, Robert Behnken
	Purpose: Third of five scheduled EVAs. Install camera, lights, and tools on Dextre; retrieve Dextre blankets; prepare Spacelab pallet for return to Earth; transfer MISSE experiments to Columbus

Linnehan and new spacewalker Robert Behnken continued assembling Dextre. They mounted Camera Light Pan Tilt Assembly (CLPA) and the tool platform and holders on the robot. Furthermore, the pair retrieved nearly all of the covers. Linnehan and Behnken prepared Dextre's Spacelab pallet to be returned to Earth with Endeavour. The pair then attempted to transfer the six MISSE experiments to Columbus, but could not due to a blocked pip pinhole.[7]

---

6. "All Aboard: The Station Goes Global," National Aeronautics and Space Administration, accessed 7 June 2014, (*http://www.nasa.gov/pdf/215905main_sts123_press_kit_b.pdf*); "Dextre, the International Space Station's Robotic Handyman," Canadian Space Agency, accessed 7 June 2014, (*http://www.asc-csa.gc.ca/eng/iss/dextre*).

7. "All Aboard: The Station Goes Global," National Aeronautics and Space Administration, accessed 5 June 2014, (*http://www.nasa.gov/pdf/215905main_sts123_press_kit_b.pdf*).

20 MARCH

2008 EVA 8	**Duration:** 6:24
World EVA 291	**Spacecraft/mission:** STS-123
U.S. EVA 192	**Crew:** Dominic Gorie, Gregory Johnson, Garrett Reisman (began tour aboard the ISS), Robert Behnken, Michael Foreman, Richard Linnehan, Takao Doi (JAXA), Leopold Eyharts (ESA, returned to Earth)
Shuttle EVA 117	**Spacewalkers:** Robert Behnken, Michael Foreman
	Purpose: Fourth of five scheduled EVAs. Replace failed RPCM and reconfigure CMG-2 power; test tile repair materials and techniques; remove Dextre thermal cover; find missing Dextre berth pin

Foreman and Behnken replaced a failed RPCM on the Z1 truss, in attempt to reconfigure power lines to CMG-2. However, they were unable to disconnect the existing electrical connections and successfully power CMG-2. The pair then experimented with Shuttle tile repair techniques and tools. They also unhinged launch locks and berthing mechanisms on Harmony. Foreman and Behnken then removed the last thermal cover on Dextre, though they could not locate a missing berth pin.[8]

22 MARCH

2008 EVA 9	**Duration:** 6:02
World EVA 292	**Spacecraft/mission:** STS-123
U.S. EVA 193	**Crew:** Dominic Gorie, Gregory Johnson, Garrett Reisman (began tour aboard the ISS), Robert Behnken, Michael Foreman, Richard Linnehan, Takao Doi (JAXA), Leopold Eyharts (ESA, returned to Earth)
Shuttle EVA 118	**Spacewalkers:** Robert Behnken, Michael Foreman
	Purpose: Fifth of five scheduled EVAs. Stow Orbiter Boom Sensor System on S1 truss; install Trundle Bearing Assembly; inspect SARJ covers; install MISSE experiments on Columbus

During the crew's last spacewalk, Behnken and Foreman temporarily stored the Orbiter Boom Sensor System (OBSS) on the S1 truss. The boom is usually connected to Canadarm2 to detect thermal damage. The pair also installed the TBA (Trundle Bearing Assembly) on the SARJ and inspected the joint covers. The TBA consists of twelve components designed to attach the SARJ between inboard and outboard truss elements. The TBA also allows the SARJ to rotate. During a previous spacewalk on 17 March, the crew was unable to install an experimental pallet on Columbus's exterior. Behnken and Foreman successfully resolved the issue and mounted the experiments.[9]

---

8. Ibid.

9. Ibid.

26 March	STS-123/Endeavour landing
8 April	Soyuz-TMA 12/ISS Expedition 17 launch
19 April	Soyuz-TMA 11/ISS Expedition 16 landing
31 May	STS-124/Discovery launch

3 JUNE

2008 EVA 10
World EVA 293
U.S. EVA 194
Shuttle EVA 119

Duration: 6:48

Spacecraft/mission: STS-124

Crew: Mark Kelly, Kenneth Ham, Karen Nyberg, Garrett Reisman (returned to Earth), Gregory Chamitoff (began tour aboard the ISS), Ronald Garan, Michael Fossum, Akihiko Hoshide (JAXA)

Spacewalkers: Michael Fossum, Ronald Garan

Purpose: First of three scheduled EVAs. Release special launch restraint of elbow camera of Shuttle robotic arm; transfer the OBSS boom from the S1 truss to the Orbiter's Payload Bay for return; prepare Japanese Experiment Module (JEM) for installation; remove berthing mechanical covers and release window locks; reinstall Trundle Bearing Assembly; confirm divot and demonstrate cleaning and lubrication techniques of the SARJ; remove SARJ launch restraints

During the three excursions, Fossum and Ronald Garan wore new spacesuit gloves with supportive thumb and index finger patches. The gloves were designed to strengthen areas where tears occurred in recent spacewalks. The supportive patches are made of the same materials already used in the palm of the glove, though it has a tighter and stronger weave called "TurtleSkin." During the initial spacewalk, Garan first unbolted the stanchions securing the OBSS to the Station S1 truss. Fossum then detached an umbilical cord that kept the OBSS powered while temporarily stowed. They then transferred it from the Station to Discovery's payload bay. It later returned to Earth with the Shuttle. Fossum then examined Harmony's CBM and confirmed it was prepared to connect with the Japanese Experiment Module Kibo. The pair readied Kibo for further assembly, and removed its contamination covers and unbolted its window locks. Garan then replaced 1 of the 12 TBAs that was not performing well. Meanwhile, Fossum examined the damaged area and retrieved debris samples on the SARJ. He also tested techniques to remove the debris and clean the race ring on the SARJ.[10]

10. "STS-124 Kibo: Hope for a New Era," National Aeronautics and Space Administration, accessed 15 June 2014, (*http://www.nasa.gov/pdf/228145main_sts124_presskit2.pdf*).

5 JUNE

2008 EVA 11	**Duration:** 7:11
World EVA 294	**Spacecraft/mission:** STS-124
U.S. EVA 195	**Crew:** Mark Kelly, Kenneth Ham, Karen Nyberg, Garrett Reisman (returned to Earth), Gregory Chamitoff (began tour aboard the ISS), Ronald Garan, Michael Fossum Akihiko Hoshide (JAXA)
Shuttle EVA 120	**Spacewalkers:** Michael Fossum, Ronald Garan
	Purpose: Second of three scheduled EVAs. Install two cameras to Kibo; remove Kibo thermal covers; prepare CBM for installation to Kibo; begin work to replace Station's nitrogen tank; install four foot restraints; retrieve failing camera

Fossum and Garan's first objective was to continue assembling Kibo. They installed two JEM Television Electronics (JTVE) cameras, one on the front end and the other on the aft end, of the laboratory. The pair then retrieved seven thermal covers off Kibo's robotic arm joints. The astronauts then prepared for the attachment of the CBM to Kibo. They removed its thermal covers, bolts, and launch locks. Their focus shifted, and they began working on the Station's Nitrogen Tank Assembly (NTA) replacement. While Garan installed thermal covers over the old nitrogen tank and disconnected its electrical and nitrogen lines, Fossum mounted four foot restraints to the work area for future spacewalks. Finally, the pair worked together to retrieve a failing camera. They repaired the camera and reinstalled it during their last spacewalk.[11]

8 JUNE

2008 EVA 12	**Duration:** 6:33
World EVA 295	**Spacecraft/mission:** STS-124
U.S. EVA 196	**Crew:** Mark Kelly, Kenneth Ham, Karen Nyberg, Garrett Reisman (returned to Earth), Gregory Chamitoff (began tour aboard the ISS), Ronald Garan, Michael Fossum, Akihiko Hoshide (JAXA)
Shuttle EVA 121	**Spacewalkers:** Michael Fossum, Ronald Garan
	Purpose: Third of three scheduled EVAs. Replace nitrogen tank on S1 truss; reinstall repaired truss camera; remove remaining Kibo launch locks; mount micrometeoroid debris shields on Kibo

On the final EVA, Fossum and Garan completed the replacement of the NTA on the S1 truss. Attached to the Canadarm2, Garan removed four bolts securing the old nitrogen tank. He then mounted a portable handle and carried the tank to the stowage platform. Meanwhile, Fossum prepared the spare nitrogen tank and mounted four bolts. Garan removed the handle from the old nitrogen tank and attached it to the new nitrogen tank. He then guided it to the S1 truss for installation. Fossum moved to the Kibo Laboratory and released launch locks on the laboratory's windows. He also removed

11. Ibid.

thermal covers off Kibo's cameras. Fossum mounted two micrometeoroid debris shields to the laboratory. The pair then reinstalled two repaired External Television Camera Group (ETVCG) cameras on the truss and concluded the spacewalk.[12]

14 June STS-124/Discovery landing

10 JULY

2008 EVA 13	**Duration:** 6:18
World EVA 296	**Spacecraft/mission:** International Space Station Expedition 17
Russian EVA 125	**Crew:** Sergei Volkov (Russian Space Agency), Gregory Chamitoff (NASA), Oleg Kononenko (Russian Space Agency)
Space Station EVA 145	**Spacewalkers:** Sergei Volkov, Oleg Kononenko
ISS EVA 40	**Purpose:** Inspect and photograph Soyuz TMA-12; retrieve faulty pyro bolt

The crew's primary objectives were to examine the recently berthed Soyuz spacecraft and recover its pyro bolt. Sergei Volkov and Oleg Kononenko exited the Pirs hatch and then, while attached to Strela, photographed the Soyuz spacecraft. They also mounted covers over its thrusters. The cosmonauts then disconnected the spacecraft's electrical cables and severed a wire to disconnect and retrieve the pyro bolt. Volkov and Kononenko then returned to Pirs with the pyro bolt, which was later returned to Earth for inspection.[13]

15 JULY

2008 EVA 14	**Duration:** 5:54
World EVA 297	**Spacecraft/mission:** International Space Station Expedition 17
Russian EVA 126	**Crew:** Sergei Volkov (Russian Space Agency), Gregory Chamitoff (NASA), Oleg Kononenko (Russian Space Agency)
Space Station EVA 146	**Spacewalkers:** Sergei Volkov, Oleg Kononenko
ISS EVA 41	**Purpose:** Install docking target for new laboratory berthing; install Vsplesk experiment; retrieve Biorisk experiment

During their second spacewalk, Volkov and Kononenko focused on new installations to the Station's exterior and preparing for the arrival of a new Russian laboratory, which was scheduled to arrive in 2010. They operated the Strela crane to install the docking target on the Zvezda module, where the new laboratory was scheduled to dock. The cosmonauts then mounted a ladder on Zvezda to allow the assembly of the Kurs antenna in 2010. Volkov and Kononenko returned to Pirs and retrieved

---

12. Ibid.

13. "Russian Spacewalkers Retrieve Soyuz Pyro Bolt," John Ira Petty, editor, accessed 20 June 2014, (*http://www.nasa.gov/mission_pages/station/expeditions/expedition17/eva20a.html*).

the Vsplesk experiment and assembled it on the wider side of Zvezda. Using high-energy particle streams, Vsplesk monitors the diffusion of Earth's elastic waves. Time permitted Volkov to complete an unscheduled task—straightening out a bent ham radio antenna. The pair concluded the spacewalk with the retrieval of the Biorisk experiment on Zvezda.[14]

25 September Shenzhou 7 launch

27 SEPTEMBER

2008 EVA 15	**Duration:** Time varied among various sources
World EVA 298	**Spacecraft/mission:** Shenzhou-7
Chinese EVA 1	**Crew:** Zhai Zhigang, Liu Boming, Jing Haipeng
	Spacewalkers: Zhai Zhigang, Liu Boming
	Purpose: Perform first Chinese EVA; wear and demonstrate Chinese-developed Feitan spacesuit; wave national flag

During their first space flight, the Shenzhou-7 crew performed China's first spacewalk. Taikonaut Jing Haipeng remained in the pressurized vehicle, while Commander Zhai Zhigang and Liu Boming performed the televised 14-minute spacewalk. Liu wore an Orlan spacesuit purchased from Russia, and Zhai wore the Chinese-made Feitan suit. At the start of the spacewalk, Zhai stated in Mandarin, "I'm feeling quite well. I greet the Chinese people and the people of the world." He clutched a handrail next to the hatch and positioned himself in view of the television cameras. Jing handed off a Chinese flag while mostly remaining in the module. Live national broadcast presented Zhai waving the flag. The historical event garnered international speculation when the Xinhua News Agency revealed details about the Shenzhou-7 launch before the spacecraft actually left Earth. The article quoted dialogue between the spacefarers during liftoff. When questioned, the news outlet simply explained it was a technical error. Furthermore, the China National Space Administration (CNSA) was criticized for allowing the vehicle to orbit within 27.96 miles (45.01 kilometers) of the ISS. Traveling approximately 4.97 miles (8 kilometers) per second, neither spacecraft had room for error.[15]

28 September Shenzhou 7 landing

12 October Soyuz-TMA 13/ISS Expedition 18 launch

23 October Soyuz-TMA 12/ISS Expedition 17 landing

14 November STS-126/Endeavour launch

---

14. "Russian Spacewalkers Outfit Station's Exterior," John Ira Petty, editor, accessed 18 June 2014, (*http://www.nasa.gov/mission_pages/station/expeditions/expedition17/eva20.html*).

15. "Chinese Astronaut Walks in Space," BBC News, accessed 19 June 2014, (*http://news.bbc.co.uk/2/hi/science/nature/7637818.stm*).

18 NOVEMBER

2008 EVA 16	**Duration:** 6:52
World EVA 299	**Spacecraft/mission:** STS-126
U.S. EVA 197	**Crew:** Christopher Ferguson, Eric Boe, Sandra Magnus (began tour aboard the ISS), Stephen Bowen, Donald Pettit, Shane Kimbrough, Heidemarie Stefanyshyn-Piper, Gregory Chamitoff (returned to Earth)
Shuttle EVA 122	
	Spacewalkers: Heidemarie Stefanyshyn-Piper, Stephen Bowen
	Purpose: First of four scheduled EVAs. Transfer empty Nitrogen Tank Assembly (NTA) to Endeavour; relocate Flex Hose Rotary Coupler to ISS; remove Kibo thermal covers; begin cleaning SARJs

The majority of the STS-126 crew's four spacewalks was spent cleaning and lubricating the SARJs. Experienced spacewalker Heidemarie Stefanyshyn-Piper performed the first three spacewalks and was designated the lead spacewalker. During the crew's first spacewalk, Stephen Bowen and Stefanyshyn-Piper retrieved an empty NTA to be returned to Earth. Stefanyshyn-Piper mounted a foot restraint on Canadarm2 and, while standing in it, guided the tank to Endeavour's payload bay. The spacewalkers then completed several smaller tasks: they retrieved a camera and closed a window flap on Harmony's CBM. They also removed a spare Flex Hose Rotary Coupler from the Lightweight Multi-Purpose Equipment Support Structure Carrier (LMC) located in the Shuttle's payload bay to the Station. The device, scheduled to be used during impending EVAs, transfers liquid ammonia to the ISS radiators. They completed the initial stages to clean the Starboard SARJ. Bowen and Stefanyshyn-Piper first wiped away debris from cleaner surfaces, and then they focused on the more damaged areas. The pair coated the outer surface with grease and scraped off the metal debris. They used a dry cloth to remove the remaining grease and metal. The spacewalkers also replaced two SARJ Trundle Bearings Assemblies.[16]

20 NOVEMBER

2008 EVA 17	**Duration:** 6:45
World EVA 300	**Spacecraft/mission:** STS-126
U.S. EVA 198	**Crew:** Christopher Ferguson, Eric Boe, Sandra Magnus (began tour aboard the ISS), Stephen Bowen, Donald Pettit, Shane Kimbrough, Heidemarie Stefanyshyn-Piper, Gregory Chamitoff (returned to Earth)
Shuttle EVA 123	
	Spacewalkers: Heidemarie Stefanyshyn-Piper, Shane Kimbrough
	Purpose: Second of four scheduled EVAs. Relocate 2 CETA carts to port side; continue cleaning SARJs; lubricate Canadarm2 grasping tool

Shane Kimbrough and Stefanyshyn-Piper's first objective was to relocate two CETA carts from the Station's starboard side to the port side. The carts were moved to prepare for the arrival of the S6 truss during the next Shuttle flight. Stefanyshyn-Piper unlocked and prepared the carts for transfer, while

16. "STS-126: Extreme Home Improvements," National Aeronautics and Space Administration, accessed 13 June 2014, (*http://www.nasa.gov/pdf/287211main_sts126_press_kit2.pdf*).

FIGURE 8. **Lubricating SARJ.** View of Shane Kimbrough as he works to relocate a Crew and Equipment Translation Aid (CETA) cart during STS-126 EVA 2. (NASA ISS018-E-009292).

Kimbrough rode Canadarm2 and individually guided them to their new location. He then lubricated the latching end effector, the robotic arm's grasping tool. The astronauts then continued polishing the SARJ and replaced two trundle bearings. They completed their tasks more quickly than expected and ended the excursion early.[17]

22 NOVEMBER

2008 EVA 18	**Duration:** 6:57
World EVA 301	**Spacecraft/mission:** STS-126
U.S. EVA 199	**Crew:** Christopher Ferguson, Eric Boe, Sandra Magnus (began tour aboard the ISS), Stephen Bowen, Donald Pettit, Shane Kimbrough, Heidemarie Stefanyshyn-Piper, Gregory Chamitoff (returned to Earth)
Shuttle EVA 124	**Spacewalkers:** Heidemarie Stefanyshyn-Piper, Stephen Bowen
	Purpose: Third of four scheduled EVAs. Replace the remaining trundle bearings; finish cleaning the SARJs

17. Ibid.

Bowen and Stefanyshyn-Piper finished cleaning the SARJs and replaced the remaining trundle bearings, except one. Using the same cleaning techniques as the two previous spacewalks, the pair spent 7 hours finalizing the maintenance of the rotary joints. They also cleaned and greased the joint's race ring.[18]

24 NOVEMBER

2008 EVA 19	**Duration:** 6:07
World EVA 302	**Spacecraft/mission:** STS-126
U.S. EVA 200	**Crew:** Christopher Ferguson, Eric Boe, Sandra Magnus (began tour aboard the ISS), Stephen Bowen, Donald Pettit, Shane Kimbrough, Heidemarie Stefanyshyn-Piper, Gregory Chamitoff (returned to Earth)
Shuttle EVA 125	**Spacewalkers:** Shane Kimbrough, Stephen Bowen
	Purpose: Fourth of four scheduled EVAs. Lubricate the SARJ; replace last trundle bearing; reinstall the covers on Kibo CBM; install one of two GPS antennae; mount camera on P1 truss; mount handrails on Kibo; photograph trusses and Mobile Transporter trailing umbilical system cables

As a preventative measure, Bowen and Kimbrough lubricated the entire surface of the SARJs again. The spacewalkers opened six of the rotary joint's covers and greased them. Ground controllers rotated the joints 180 degrees to allow the grease to spread and expose any unlubricated areas. While Kimbrough greased any remaining surfaces, Bowen worked on the experiments attached to the Kibo Laboratory's exterior. He also reinstalled the CBM cover that he removed during his first spacewalk. Afterward, Bowen wrapped the Canadarm2's tabs around a cable to prevent them from blocking the robotic arm's camera. Bowen returned to Kibo and mounted three handrails and a GPS antenna. The spacewalkers photographed the trusses and the mobile transporter's cables and finalized the excursion.[19]

30 November STS-126/Endeavour landing

22 DECEMBER

2008 EVA 20	**Duration:** 5:38
World EVA 303	**Spacecraft/mission:** International Space Station Expedition 18
Russian EVA 127/ U.S. EVA 201	**Crew:** Michael Fincke (NASA), Yuri Lonchakov (Russian Space Agency), Sandra Magnus (NASA), Koichi Wakata (JAXA)
Space Station EVA 147	**Spacewalkers:** Yuri Lonchakov, Michael Fincke
ISS EVA 42	**Purpose:** Install Langmuir probe to aid Soyuz reentry separation analysis; retrieve Biorisk can; install Expose-R and Impulse experiment

18. Ibid.

19. Ibid.

Yuri Lonchakov and Fincke set out on their spacewalk at 7:51 p.m. EST on Monday, 22 December and returned to ISS's airlock at 1:29 a.m. the following day. They installed a Langmuir probe on Pirs to measure electromagnetic energy to determine its effects on pyrotechnical separation bolts on Soyuz. Fincke removed Biorisk experiment canister from outside of Pirs, and he and Lonchakov put it back inside the airlock. They then installed the two new experiments, Expose-R and Impulse, on Zvezda. Conducted by ESA, Expose-R, like the Biorisk experiment, exposed biological substances to outer space. Impulse measured disturbances in the ionosphere surrounding ISS. They were able to connect both experiments in spite of having trouble with some stubborn cables. Moscow flight controllers relayed that the Impulse experiment was functional but they did not receive telemetry from Expose-R. Therefore, they ordered the pair to disconnect Expose-R and bring it back with them to Pirs.[20]

---

20. "Station Spacewalkers Install Experiments," Amiko Kauderer, editor, accessed 14 June 2014, (*http://www.nasa.gov/mission_pages/station/expeditions/expedition18/eva21.html*).

2009

10 MARCH

2009 EVA 1	**Duration:** 4:49
World EVA 304	**Spacecraft/mission:** International Space Station Expedition 18
Russian EVA 128/ U.S. EVA 202	**Crew:** Michael Fincke (NASA), Yuri Lonchakov (Russian Space Agency), Sandra Magnus (NASA)
Space Station EVA 148	**Spacewalkers:** Yuri Lonchakov, Michael Fincke
ISS EVA 43	**Purpose:** Cut off excess length of fabric straps near DC1 antennas and docking target; reinstall Expose-R experiment; adjust orientation of SKK experiment panel; close thermal insulation flap over electrical connector panel; photograph external hardware

Fincke and Lonchakov brought the Expose-R experiment back to Zvezda, re-attempting to install it. This time the installation was a success. They also removed tape on Pirs' docking target and docking compartment to ensure that it would not interfere with the arrival of Soyuz or other spacecraft. The pair concluded their walk by taking photographs of the exterior of the Russian portion of ISS. This was to assess the current condition of the Station's surface and its mechanical components after having been exposed to the harsh environment of space for 10 years.[1]

15 March STS-119/Discovery launch

19 MARCH

2009 EVA 2	**Duration:** 6:07
World EVA 305	**Spacecraft/mission:** STS-119
U.S. EVA 203 **Shuttle EVA 126**	**Crew:** Lee Archambault, Dominic Antonelli, Joseph Acaba, Steven Swanson, Richard Arnold, John Phillips, Steven Swanson, Koichi Wakata (JAXA, began tour aboard the ISS), Sandra Magnus (returned to Earth)
	Spacewalkers: Steven Swanson, Richard Arnold
	Purpose: First of three scheduled EVAs. Assemble final truss segment

Richard Arnold and Swanson bolted the final truss segment to the starboard side of ISS. Phillips and Wakata used Canadarm2 to pull the 15.5-short-tons (14.06-metric-tons), 45-feet (13.71-meters) long segment out of the Shuttle's payload bay and bring it into position, where Arnold and Swanson

1. "Station Spacewalkers Install Experiment, Probe," Amiko Kauderer, editor, accessed 14 June 2014, (*http://www.nasa.gov/mission_pages/station/expeditions/expedition18/eva22.html*).

promptly bolted it in place and connected to it the Station's data and power cables. The new addition to the truss brought it to a total of 335 feet (102.12 meters).[2]

21 MARCH

2009 EVA 3	**Duration:** 6:30
World EVA 306	**Spacecraft/mission:** STS-119
U.S. EVA 204	**Crew:** Lee Archambault, Dominic Antonelli, Joseph Acaba, Steven Swanson, Richard Arnold, John Phillips, Steven Swanson, Koichi Wakata (JAXA, began tour aboard the ISS), Sandra Magnus (returned to Earth)
Shuttle EVA 127	**Spacewalkers:** Steven Swanson, Joseph Acaba
	Purpose: Second of three scheduled EVAs. Loosen restraints of P6 batteries for future replacement; install JEM GPS antenna; take infrared images of P1 and S1 radiators

Joseph Acaba and Swanson prepared a worksite outside of the Station for the installation of new batteries that would be brought up by Endeavour on the STS-127 mission in July 2009. They then installed a GPS antenna on the pressurized logistics module outside the Kibo laboratory. Acaba and Swanson were scheduled to install a UCCAS, but they encountered a problem with a misaligned bracket, and it proved to be impossible to reposition it with the tools available. They were forced to postpone the installation, and they instead took infrared photos of radiators on P1 and S1.[3]

23 MARCH

2009 EVA 4	**Duration:** 6:27
World EVA 307	**Spacecraft/mission:** STS-119
U.S. EVA 205	**Crew:** Lee Archambault, Dominic Antonelli, Joseph Acaba, Steven Swanson, Richard Arnold, John Phillips, Steven Swanson, Koichi Wakata (JAXA, began tour aboard the ISS), Sandra Magnus (returned to Earth)
Shuttle EVA 128	**Spacewalkers:** Richard Arnold, Joseph Acaba
	Purpose: Third of three scheduled EVAs. Relocate one CETA cart from port to starboard side of mobile transporter; replace CETA cart coupler; lubricate robotic arm end effector snares; reconfigure electrical connections of truss; attach bolts and free clamps of flex hose rotary coupler

2. "NASA's STS-119 Mission: Boosting the Station Power," Jeanne Ryba, editor, accessed 14 June 2014, (*http://www.nasa.gov/mission_pages/shuttle/behindscenes/119_overview.html*); "STATUS REPORT: STS-119-09," Amiko Kauderer, editor, accessed 15 June 2014, (*http://www.nasa.gov/mission_pages/shuttle/shuttlemissions/sts119/news/STS-119-09.html*).

3. "NASA's STS-119 Mission: Boosting the Station Power," Jeanne Ryba, editor, accessed 14 June 2014, (*http://www.nasa.gov/mission_pages/shuttle/behindscenes/119_overview.html*); "STATUS REPORT: STS-119-13," Amiko Kauderer, editor, accessed 14 June 2014, (*http://www.nasa.gov/mission_pages/shuttle/shuttlemissions/sts119/news/STS-119-13.html*).

In the mission's final spacewalk, Acaba and Arnold teamed up this time to move one of the CETA carts from one side of the Mobile Transporter to the other. The astronauts moved the cart to facilitate attachment of the Mobile Transporters to the Japanese science laboratory which would take place during the STS-127 mission in July 2009, as well as other future tasks. The team then attempted to deploy one of the Station's spare equipment platforms, but they could not due to a stuck mechanism. They were also supposed to work on a similarly structured payload attach system on the other side of the truss, but due to the problem with the mechanism on the spare equipment platform and a similar problem that Swanson and Acaba encountered on the previous EVA, Mission Control ordered them to forgo the work in case it was a broader problem. Finally, Acaba and Arnold lubricated the end effector capture snares on Canadarm2 to prevent it from snagging or fitting too snugly into its groove as it retracts back into the latching mechanism.[4]

26 March Soyuz-TMA 14/ISS Expedition 19 launch

28 March STS-119/Discovery landing

8 April Soyuz-TMA 13/ISS Expedition 18 landing

11 May STS-125/Atlantis launch

14 MAY

2009 EVA 5	**Duration:** 7:20
World EVA 308	**Spacecraft/mission:** STS-125
U.S. EVA 206	**Crew:** Scott Altman, Gregory Johnson, Michael Massimino, Michael Good, K. Megan McArthur, John Grunsfeld, Andrew Feustel
Shuttle EVA 129	**Spacewalkers:** John Grunsfeld, Andrew Feustel
	Purpose: Hubble Servicing Mission 4, the first of five scheduled EVAs

The STS-125 crew's schedule to perform five spacewalks during five consecutive days was demanding and historic. It was the last Hubble Servicing Mission, and it notably upgraded the equipment. The crew installed updated equipment that improved data connections between the telescope and engineers on Earth. Fortunately, the Hubble Space Telescope was passing over Kennedy Space Center when Atlantis launched, allowing the Shuttle to capture the Hubble by the second flight day. Grunsfeld and Massimino—both experienced spacewalkers who had previously repaired the Hubble—paired up with first-time spacewalkers Andrew Feustel and Michael Good. Though each excursion was scheduled for 6 hours, most actually required 7 to 8 hours. At the start of the first EVA, Grunsfeld assembled a foot restraint to allow Feustel to replace the Wide Field Camera 2 with the 900-pound (408.24-kilogram) Wide Field Camera 3. The foot restraints also shielded the solar arrays from vibrations caused by Feustel's movements. The new camera is capable of larger and more focused images

4. "NASA's STS-119 Mission: Boosting the Station Power," Jeanne Ryba, editor, accessed 14 June 2014, (*http://www.nasa.gov/mission_pages/shuttle/behindscenes/119_overview.html*); "STATUS REPORT: STS-119-17," Amiko Kauderer, editor, accessed 14 June 2014, (*http://www.nasa.gov/mission_pages/shuttle/shuttlemissions/sts119/news/STS-119-17.html*).

than its predecessor. While Feustel positioned himself to replace the camera, Grunsfeld mounted a protective blanket over Hubble's low-gain antenna. They then permanently stored the old camera in the Wide-field Scientific Instrument Protective Enclosure and installed the new camera. The astronauts also replaced the failed Science Instrument Command and Data Handling Unit, a computer that receives commands and transmits data to Earth. Then, on Hubble's aft bulkhead, they installed the Soft Capture Mechanism, a circular device that eases future capture and release of the Hubble by pilotless spacecraft. Furthermore, they installed two of three latch kits on the doors to assist following spacewalks. The kits quicken the opening and closing of Hubble's doors.[5]

15 MAY

2009 EVA 6	**Duration:** 7:56
World EVA 309	**Spacecraft/mission:** STS-125
U.S. EVA 207	**Crew:** Scott Altman, Gregory Johnson, Michael Massimino, Michael Good, K. Megan McArthur, John Grunsfeld, Andrew Feustel
Shuttle EVA 130	**Spacewalkers:** Michael Massimino, Michael Good
	Purpose: Hubble Servicing Mission 4, the second of five scheduled EVAs

For the majority of the spacewalk, Good and Massimino replaced three Rate Sensor Units. Massimino rode Atlantis' robotic arm to retrieve the units and store them in the Shuttle's cargo bay. Good assisted Massimino and maneuvered between the worksite and the cargo bay for each Rate Sensor Unit. The pair then replaced one of the telescope's two Battery Module Units. It required Good to unscrew 14 bolts to remove the 460-pound (208.65-kilogram) used battery and install its replacement. The second battery replacement was completed during the crew's final spacewalk.[6]

16 MAY

2009 EVA 7	**Duration:** 6:36
World EVA 310	**Spacecraft/mission:** STS-125
U.S. EVA 208	**Crew:** Scott Altman, Gregory Johnson, Michael Massimino, Michael Good, K. Megan McArthur, John Grunsfeld, Andrew Feustel
Shuttle EVA 131	**Spacewalkers:** John Grunsfeld, Andrew Feustel
	Purpose: Hubble Servicing Mission 4, the third of five scheduled EVAs

---

5. "Hubble Space Telescope Servicing Mission 4: The Soft Capture and Rendezvous System," Lori Tyahla, editor, accessed 15 June 2014, (*http://www.nasa.gov/mission_pages/hubble/servicing/SM4/main/SCRS_FS_HTML.html*); "Mission Accomplished: Leaving Hubble Better Than Ever," Cheryl L. Mansfield, accessed 15 June 2014, (*http://www.nasa.gov/mission_pages/shuttle/shuttlemissions/sts125/launch/125_overview.html*); "STS-125: The Final Visit to Hubble," National Aeronautics and Space Administration, accessed 20 June 2014, (*http://www.nasa.gov/pdf/331922main_sts125_presskit_050609.pdf*).

6. "Mission Accomplished: Leaving Hubble Better Than Ever," Cheryl L. Mansfield, accessed 15 June 2014, (*http://www.nasa.gov/mission_pages/shuttle/shuttlemissions/sts125/launch/125_overview.html*); "STS-125: The Final Visit to Hubble," National Aeronautics and Space Administration, accessed 20 June 2014, (*http://www.nasa.gov/pdf/331922main_sts125_presskit_050609.pdf*).

Feustel and Grunsfeld replaced the Corrective Optics Space Telescope Axial Replacement (COSTAR) with the new Cosmic Origins Spectrograph. The removal of COSTAR required Feustel to ride the robotic arm and unhook four security connectors, sever a ground strap, and unbolt two latches. They stored the old instrument in Atlantis's cargo bay and retrieved its replacement. Again, Feustel was attached to the arm. The pair secured the spectrograph with four connectors and a ground strap. Next, they also repaired the failed ACS by installing a new power supply and replacing four detector circuit boards.[7]

17 MAY

2009 EVA 8	**Duration:** 8:02
World EVA 311	**Spacecraft/mission:** STS-125
U.S. EVA 209	**Crew:** Scott Altman, Gregory Johnson, Michael Massimino, Michael Good, K. Megan McArthur, John Grunsfeld, Andrew Feustel
Shuttle EVA 132	**Spacewalkers:** Michael Massimino, Michael Good
	Purpose: Hubble Servicing Mission 4, the fourth of five scheduled EVAs

Massimino and Good replaced a failed power supply board in the Space Telescope Imaging Spectrograph (STIS). To reach the old power supply, the spacewalkers removed a cover plate and a handrail—both required tools specifically designed for the mission. The cover alone required the removal of 111 fasteners, which also required a second fastener capture plate to prevent them from floating away. Next, they extracted the old power supply board with another tool specifically designed for the mission. Lead Flight Director Tony Ceccacci compared the complexity and dexterous demands of the repair to brain surgery. The repair required more time than expected, causing the installation of two protective thermal insulation panels to be delayed.[8]

18 MAY

2009 EVA 9	**Duration:** 7:02
World EVA 312	**Spacecraft/mission:** STS-125
U.S. EVA 210	**Crew:** Scott Altman, Gregory Johnson, Michael Massimino, Michael Good, K. Megan McArthur, John Grunsfeld, Andrew Feustel
Shuttle EVA 133	**Spacewalkers:** John Grunsfeld, Andrew Feustel
	Purpose: Hubble Servicing Mission 4, the fifth of five scheduled EVAs

---

7. "Mission Accomplished: Leaving Hubble Better Than Ever," Cheryl L. Mansfield, accessed 15 June 2014, (*http://www.nasa.gov/mission_pages/shuttle/shuttlemissions/sts125/launch/125_overview.html*); "STS-125: The Final Visit to Hubble," National Aeronautics and Space Administration, accessed 20 June 2014, (*http://www.nasa.gov/pdf/331922main_sts125_presskit_050609.pdf*).

8. "Mission Accomplished: Leaving Hubble Better Than Ever," Cheryl L. Mansfield, accessed 15 June 2014, (*http://www.nasa.gov/mission_pages/shuttle/shuttlemissions/sts125/launch/125_overview.html*); "STS-125: The Final Visit," Amiko Kauderer, editor, accessed 23 June 2014, (*http://www.nasa.gov/mission_pages/shuttle/shuttlemissions/sts125/main/overview.html*); "STS-125: The Final Visit to Hubble," National Aeronautics and Space Administration, accessed 20 June 2014, (*http://www.nasa.gov/pdf/331922main_sts125_presskit_050609.pdf*).

Massimino and McArthur first replaced the second Battery Module Unit, a task left incomplete during the pair's last spacewalk. Next, Grunsfeld rode the robotic arm to retrieve the old Fine Guidance Sensor, and delivered it to Atlantis's cargo bay. He guided the new sensor back to the worksite and successfully linked its nine power connectors. Finally, the spacewalkers installed three New Outer Layer Blankets to the telescope's three bays. During cleanup, Feustel and Grunsfeld inadvertently broke the tip of a low-gain antenna. They reinstalled its cover to shield the antenna permanently.[9]

24 May STS-125/Atlantis landing

27 May Soyuz-TMA 15/ISS Expeditions 20 and 21 launch

5 JUNE

2009 EVA 10	**Duration:** 4:54
World EVA 313	**Spacecraft/mission:** International Space Station Expedition 20
Russian EVA 129/ U.S. EVA 211	**Crew:** Gennady Padalka (Russian Space Agency), Michael Barratt (NASA), Robert Thirsk (Canadian Space Agency), Roman Romanenko (Russian Space Agency), Koichi Wakata (JAXA), Frank De Winne (European Space Agency)
Space Station EVA 149	
ISS EVA 44	**Spacewalkers:** Gennady Padalka, Michael Barratt
	Purpose: Install two docking navigation antennas on Zenith port of service module

Padalka and Michael Barratt teamed up to work on the Zvezda module in preparation for the addition of the new Russian Mini-Research Module-2 (MRM-2), which would be installed later in 2009, and would dock on the Zvezda's zenith port and become an additional port for the docking of Russian vehicles. On Zvezda, Padalka and Barratt installed docking antennas, a docking target, and electrical connectors specifically for the Kurs automated docking equipment. Once they completed the installation of the new equipment, Barratt rode the Strela crane, from which he photographed the new antennas. The EVA was delayed slightly and began at 3:52 a.m. because Russian teams on the ground said data they had received from the spacewalkers' spacesuits revealed that they contained high levels of carbon dioxide. After the EVA was complete, Barratt and Padalka appeared unharmed, and they said that they "felt fine."[10]

9. "Mission Accomplished: Leaving Hubble Better Than Ever," Cheryl L. Mansfield, accessed 15 June 2014, (*http://www.nasa.gov/mission_pages/shuttle/shuttlemissions/sts125/launch/125_overview.html*); "STS-125: The Final Visit to Hubble," National Aeronautics and Space Administration, accessed 20 June 2014, (*http://www.nasa.gov/pdf/331922main_sts125_presskit_050609.pdf*).

10. "Russian Spacewalk to Prepare for New Module Complete," Amiko Kauderer, editor, accessed 20 June 2014, (*http://www.nasa.gov/mission_pages/station/expeditions/expedition20/eva22.html*).

10 JUNE

2009 EVA 11	**Duration:** 0:12
World EVA 314	**Spacecraft/mission:** International Space Station Expedition 20
Russian EVA 130/ U.S. EVA 212	**Crew:** Gennady Padalka (Russian Space Agency), Michael Barratt (NASA), Robert Thirsk (Canadian Space Agency), Roman Romanenko (Russian Space Agency), Koichi Wakata (JAXA), Frank De Winne (European Space Agency)
Space Station EVA 150	
ISS EVA 45	**Spacewalkers:** Gennady Padalka, Michael Barratt
	Purpose: Replaced Zvezda hatch with a docking cone

At only 12 minutes, this was the shortest spacewalk ever performed. Taking place entirely inside the Zvezda Service Module, with the spacewalkers' suits still attached to umbilicals, it has been called an "internal" spacewalk, but is nonetheless considered a spacewalk because Barratt and Padalka worked in a depressurized space. The EVA crew replaced a hatch with a docking point. With Barratt and Padalka quickly completing the task, Zvezda became ready for the docking of MRM-2.[11]

15 July STS-127/Endeavour launch

18 JULY

2009 EVA 12	**Duration:** 5:32
World EVA 315	**Spacecraft/mission:** STS-127
U.S. EVA 213	**Crew:** Mark Polansky, Douglas Hurley, David Wolf, Christopher Cassidy, Julie Payette, Thomas Marshburn, Timothy Kopra (began tour aboard the ISS), Koichi Wakata (JAXA, returned to Earth)
Shuttle EVA 134	**Spacewalkers:** David Wolf, Timothy Kopra
	Purpose: First of five scheduled EVAs. Jettison Kibo's CBM; remove Kibo launch heater cable; reposition Kibo robotic arm ground tab; install CBM blankets on Unity and Harmony; reconfigure CETA cart brake and foot restraints; install UCCAS and the payload attachment system; remove cover from the Monitor of All-sky X-ray Image

The STS-127 crew was the third to perform five spacewalks. Their main objective was to complete the installation of the Kibo Laboratory, in addition to finalizing uncompleted tasks from the STS-119 mission. Experienced spacewalker Wolf led the five EVAs and guided first-time spacewalkers Christopher Cassidy, Timothy Kopra, and Thomas Marshburn. During the first walk, Wolf and Kopra concluded preparations to install Kibo permanently. The actual installation was later completed robotically. Wolf jettisoned the insulating blankets covering the CBM on Kibo and removed the laboratory's

11. "Expedition 19 and 20: Full Partners," National Aeronautics and Space Administration, accessed 21 June 2014, (*http://www.nasa.gov/pdf/320539main_Expedition_19_20_Press_Kit.pdf*); "Russian 'Internal' Spacewalk Complete," Amiko Kauderer, editor, accessed 21 June 2014, (*http://www.nasa.gov/mission_pages/station/expeditions/expedition20/eva23.html*).

launch heater cable. He also repositioned grounding tabs that obscured the robotic arm's camera view. Meanwhile, in Endeavour's cargo bay, Kopra released the exposed facility's four fasteners, insulation covers, and temporary power cable. The exposed facility is a platform that houses up to 10 experiments on Kibo's exterior. Kopra confirmed the grasp of the Shuttle's robotic arm on the exposed facility, allowing it to pass the device to Canadarm2. The spacewalkers then focused on retrieving the Station's new spare equipment, delivered in Endeavour's cargo bay. After removing a contamination cover from the Monitor of All-sky X-ray Image (MAXI) in the bay, Kopra made his way to the Harmony node and secured a CBM cover. Next, he retrieved tools on the zenith truss and removed the cover on the Unity node's CBM. Meanwhile, Wolf moved to the Japanese laboratory and unscrewed the ammonia tank's four bolts, preparing for the succeeding EVA's tasks. He then made his way to the CETA cart, on the Station's port side, and repositioned its brake handles and foot restraints to avoid an interference with the SARJ. Next, Wolf and Kopra worked together to assemble the UCCAS and the Payload Attachment System (PAS). Though they have different names, the two are nearly identical systems. The devices store cargo on the trusses.[12]

20 JULY

2009 EVA 13
World EVA 316
U.S. EVA 214
Shuttle EVA 135

Duration: 6:53

Spacecraft/mission: STS-127

Crew: Mark Polansky, Douglas Hurley, David Wolf, Christopher Cassidy, Julie Payette, Thomas Marshburn, Timothy Kopra (began tour aboard the ISS), Koichi Wakata (JAXA, returned to Earth)

Spacewalkers: David Wolf, Thomas Marshburn

Purpose: Second of five scheduled EVAs. Install space-to-ground antenna; install space pump module, install linear drive unit; assemble camera on exposed facility

Prior to the start of the EVA, the crew remotely unpacked the spare Station equipment with Endeavour's robotic arm. Wolf rode Canadarm2 and separately transferred three large spare devices from the Shuttle payload bay to the P3 truss's external stowage platform, and Marshburn would then install the equipment. Transferring and installing the space-to-ground antenna, the space pump module, and the linear drive unit required 3.5 hours. For the remainder of the EVA, the spacewalkers assembled cameras on Kibo's exposed facility.[13]

12. "STS-126: A Porch in Space," National Aeronautics and Space Administration, 10 June 2014, (*http://www.nasa.gov/pdf/358018main_sts127_presskit.pdf*); "STS-126: Extreme Home Improvements," National Aeronautics and Space Administration, accessed 5 June 2015, (*http://www.nasa.gov/pdf/287211main_sts126_press_kit2.pdf*).

13. "Completing Kibo: STS-127 Marks New Era for Science," Ann Heiney, accessed 10 June 2014, (*http://www.nasa.gov/mission_pages/shuttle/shuttlemissions/sts127/launch/127_overview.html*); "STS-126: A Porch in Space," National Aeronautics and Space Administration, 10 June 2014, (*http://www.nasa.gov/pdf/358018main_sts127_presskit.pdf*).

22 JULY

2009 EVA 14	**Duration:** 5:59
World EVA 317	**Spacecraft/mission:** STS-127
U.S. EVA 215	**Crew:** Mark Polansky, Douglas Hurley, David Wolf, Christopher Cassidy, Julie Payette, Thomas Marshburn, Timothy Kopra (began tour aboard the ISS), Koichi Wakata (JAXA, returned to Earth)
Shuttle EVA 136	**Spacewalkers:** David Wolf, Christopher Cassidy
	Purpose: Third of five scheduled EVAs. Install exposed section on Kibo; install four batteries on P6 truss; stow old batteries in Shuttle cargo bay

At the start of the excursion, Wolf relocated the tools and handrails from Harmony to Columbus, their new worksite. Meanwhile, Cassidy readied three uninstalled storage apparatuses for the exposed portion of the facility, called "exposed section." While in Endeavour's cargo bay, he removed their thermal insulation. From within the Shuttle, Julie Payette later installed the exposed sections using the robotic arm. Next, Cassidy retrieved two covers from the inter orbit communication system and released its antenna-holding device. Though they were scheduled to replace four of the six batteries in the P6 truss, Cassidy's high work rate caused an unexpected increase of carbon dioxide in his suit and forced the walk to end early. They changed two batteries and delayed replacing the remaining four until the following excursion.[14]

24 JULY

2009 EVA 15	**Duration:** 7:12
World EVA 318	**Spacecraft/mission:** STS-127
U.S. EVA 216	**Crew:** Mark Polansky, Douglas Hurley, David Wolf, Christopher Cassidy, Julie Payette, Thomas Marshburn, Timothy Kopra (began tour aboard the ISS), Koichi Wakata (JAXA, returned to Earth)
Shuttle EVA 137	**Spacewalkers:** Thomas Marshburn, Christopher Cassidy
	Purpose: Fourth of five scheduled EVAs. Replace four batteries on P6 truss; install exposed facility camera

Cassidy and Marshburn replaced four batteries on the P6 truss and finalized the task from the previous spacewalk. They stowed the used batteries in Endeavour's payload bay for return to Earth. The pair delayed assembling the exposed facility camera, a scheduled task, until the succeeding walk.[15]

---

14. "Completing Kibo: STS-127 Marks New Era for Science," Ann Heiney, accessed 10 June 2014, (*http://www.nasa.gov/mission_pages/shuttle/shuttlemissions/sts127/launch/127_overview.html*); "STS-126: A Porch in Space," National Aeronautics and Space Administration, 10 June 2014, (*http://www.nasa.gov/pdf/358018main_sts127_presskit.pdf*); "What is the Japanese Experiment Module, 'Kibo'?," Japanese Aerospace Exploration Agency, accessed 10 June 2014, (*http://iss.jaxa.jp/kids/en/station/05.html*).

15. "STS-126: A Porch in Space," National Aeronautics and Space Administration, 10 June 2014, (*http://www.nasa.gov/pdf/358018main_sts127_presskit.pdf*); "STS-127: Mission Archive," William Harwood, accessed June 10, 2014, (*http://www.cbsnews.com/network/news/space/127/STS-127_Archive.html*).

27 JULY

2009 EVA 16	**Duration:** 4:54
World EVA 319	**Spacecraft/mission:** STS-127
U.S. EVA 217	**Crew:** Mark Polansky, Douglas Hurley, David Wolf, Christopher Cassidy, Julie Payette, Thomas Marshburn, Timothy Kopra (began tour aboard the ISS), Koichi Wakata (JAXA, returned to Earth)
Shuttle EVA 138	**Spacewalkers:** Thomas Marshburn, Christopher Cassidy
	Purpose: Fifth of five scheduled EVAs. Secure thermal covers over Dextre's orbital replacement unit tool; install two external video cameras to Kibo; reconfigure electrical connectors on the Z1 truss; install two external facility video cameras; install handrail and foot restraint on Kibo

Cassidy and Marshburn focused on completing get-ahead tasks for the upcoming STS-128 mission. Marshburn secured two thermal insulations on Dextre's wrist joints, known as the "orbital replacement unit tool." Meanwhile, Cassidy reconfigured two electrical connectors on the Z1 truss's patch panel to restore power to the CMG. They also mounted two external video cameras on the Japanese laboratory's exposed facility—a task initially scheduled for the previous spacewalk. Additionally, on Kibo, they mounted a handrail, a foot restraint, and a gap spanner, which is a fabric strap that can be used to route fluid umbilicals, act as a handrail, or anchor for tethers. The CO_2 scrubber abruptly failed and forced the spacewalkers to desert the installation of two PAS and WETA to the S3 truss.[16]

31 July STS-127/Endeavour landing

28 August STS-128/Discovery launch

1 SEPTEMBER

2009 EVA 17	**Duration:** 6:35
World EVA 320	**Spacecraft/mission:** STS-128
U.S. EVA 218	**Crew:** Frederick Sturckow, Kevin Ford, Jose Hernandez, Danny Olivas, Nicole Stott (began tour aboard the ISS), Patrick Forrester, Timothy Kopra (returned to Earth), Christor Fuglesang (European Space Agency)
Shuttle EVA 139	**Spacewalkers:** Danny Olivas, Nicole Stott
	Purpose: First of three scheduled EVAs. Remove empty ammonia tank; retrieve MISSE and EuTEF experiments

During her first EVA, Nicole Stott was guided by veteran spacewalker Olivas. They performed maintenance tasks on the Station. The spacewalkers first removed a depleted ammonia tank from the P1 truss. They disconnected its two ammonia, two nitrogen, and two electrical lines. With the assistance of the Canadarm2, Olivas and Stott temporarily stored the empty tank until the succeeding spacewalk. The

16. Ibid.

pair then retrieved MISSE and EuTEF experiments from the ESA's Columbus laboratory and returned them to the Shuttle.[17]

3 SEPTEMBER

2009 EVA 18	**Duration:** 6:39
World EVA 321	**Spacecraft/mission:** STS-128
ESA EVA 10/U.S. EVA 219	**Crew:** Frederick Sturckow, Kevin Ford, Jose Hernandez, Danny Olivas, Nicole Stott (began tour aboard the ISS), Patrick Forrester, Timothy Kopra (returned to Earth), Christer Fuglesang (European Space Agency)
Shuttle EVA 140	**Spacewalkers:** Danny Olivas, Christer Fuglesang
	Purpose: Second of three scheduled EVAs. Install new ammonia tank on P1 truss; stow depleted ammonia tank in Discovery cargo bay

The second spacewalk centered on the storing of the depleted ammonia tank and installing a new tank on the P1 truss. While Olivas detached the insulation covers on the new tank, Fuglesang positioned himself on the Canadarm2 foot restraints. Next, they unscrewed four bolts securing the new tank to Discovery. Astronauts Kevin Ford and Stott operated the Canadarm2 and drove Fuglesang and the tanks to the installation site. The spacewalkers fastened the new tank with four bolts and connected two power cables and four fluid lines. With the Canadarm2, Fuglesang and Olivas stowed the old tank and secured it to the Shuttle's cargo bay. It returned to Earth with Discovery and was later refilled and returned to the Station.[18]

5 SEPTEMBER

2009 EVA 19	**Duration:** 7:01
World EVA 322	**Spacecraft/mission:** STS-128
ESA EVA 11/U.S. EVA 220	**Crew:** Frederick Sturckow, Kevin Ford, Jose Hernandez, Danny Olivas, Nicole Stott (began tour aboard the ISS), Patrick Forrester, Timothy Kopra (returned to Earth), Christer Fuglesang (European Space Agency)
Shuttle EVA 141	**Spacewalkers:** Danny Olivas, Christer Fuglesang
	Purpose: Third of three scheduled EVAs. Install cargo attachment system on S3 truss; replace CMG-2; prepare heater lines on PMA-3 for Tranquility docking; replace faulty RPCM

During the last EVA by STS-128, Fuglesang and Olivas focused on completing unfinished tasks from previous missions, in addition to preparing the Station for future spacewalks and the berthing of the

17. "STS-128: Outfits Station for New Science," Steve Siceloff, accessed 15 June 2014, (*http://www.nasa.gov/mission_pages/shuttle/shuttlemissions/sts128/launch/128_overview.html*); "STS-128 Racking Up New Science," National Aeronautics and Space Administration, accessed 15 June 2014, (*http://www.nasa.gov/pdf/379392main_STS-128_Press_Kit.pdf*).

18. Ibid.

Node 3, commonly known as the Tranquility module. The pair successfully deployed a cargo attachment system on the S3 truss, which the STS-127 crew did not have time to complete due to a setback caused by the CO_2 scrubber. Next, they replaced the failing CMG-2 and temporarily stowed it nearby. The spacewalkers then separated and undertook individual tasks. Olivas readied the heater lines to keep PMA-3 warm when Tranquility and Unity later coupled. Fuglesang moved to the S0 truss and simply removed a bolt to retrieve the faulty RPCM. He slid the new device in place and reinstalled the same screw. Additionally, he installed one of two insulation cables on the truss that later connected with the new node. Fuglesang was unable to connect the other cable. At the end of the spacewalk, Olivas detached a damaged slidewire on Unity, another necessary task to dock Tranquility. By the end of the excursion, spacewalkers spent more than 830 hours building the Station.[19]

11 September STS-128/Discovery landing

30 September Soyuz-TMA 16/ISS Expeditions 21 and 22 launch

11 October Soyuz-TMA 14/ISS Expedition 20 landing

16 November STS-129/Atlantis launch

19 NOVEMBER

2009 EVA 20	**Duration:** 6:37
World EVA 323	**Spacecraft/mission:** STS-129
U.S. EVA 221	**Crew:** Charles Hobaugh, Barry Wilmore, Michael Foreman, Leland Melvin, Robert Satcher, Nicole Stott (returned to Earth), Randolph Bresnik
Shuttle EVA 142	**Spacewalkers:** Michael Foreman, Robert Satcher
	Purpose: First of three scheduled EVAs. Install S-band antenna; replace handrail on Unity with line bracket; lubricate mobile transporter; lubricate Kibo robot arms

The weight and volume of this mission's cargo could only be delivered by Atlantis, the only Shuttle able to launch with a payload so large. The STS-129 mission was dedicated to restocking various instruments and preparing for the arrival of Tranquility. Most parts were installed robotically, but some required manual assembly. Experienced spacewalker Foreman guided Robert Satcher during his first EVA. Satcher climbed onto a Canadarm2 foot restraint and guided the S-band antenna to the Z1 truss, where Foreman then installed it. While he waited for Satcher, Foreman installed an ingress aid in the airlock and stowed another in the CETA cart. The spacewalkers then separated and undertook individual tasks. Foreman secured cables for future space-to-ground antennae, replaced a handrail with a bracket later used to hold Tranquility's ammonia cable, repositioned Unity's cable connectors,

19. Ibid.

and secured its tool tethers. Meanwhile, Satchel used a grease gun to lubricate Kibo's robotic arms, the mobile transporter's tethers, and the attached equipment.[20]

21 NOVEMBER

2009 EVA 21	**Duration:** 6:08
World EVA 324	**Spacecraft/mission:** STS-129
U.S. EVA 222	**Crew:** Charles Hobaugh, Barry Wilmore, Michael Foreman, Leland Melvin, Robert Satcher, Nicole Stott (returned to Earth), Randolph Bresnik
Shuttle EVA 143	**Spacewalkers:** Michael Foreman, Randolph Bresnik
	Purpose: Second of three scheduled EVAs. Install the Grappling Adaptor to On-Orbit Railing (GATOR); install two PAS and video camera on S3 truss; install WETA on S3 truss; relocate FPMU from S1 truss to P1 truss; photograph faulty seals on Harmony CBM

Randolph Bresnik and Foreman collaborated to complete the scheduled tasks. They first undertook the GATOR installation onto Columbus. Astrium Bremen in Germany designed GATOR, a device comprised of two components: a ham radio antenna and a ship-tracking antenna system called "Automatic Identification System." The United States Coast Guard's Vessel Traffic Services uses the latter to exchange data between ships. Bresnik and Foreman then relocated the FPMU from the S1 truss to the P1 truss. Next, the spacewalkers deployed the remaining two PAS and a wireless camera on the S3 truss. Once mounted on the truss, PAS houses spare parts delivered by future missions. Bresnik and Foreman installed the Station's second WETA on the S3 truss and concluded the walk.[21]

23 NOVEMBER

2009 EVA 22	**Duration:** 5:42
World EVA 325	**Spacecraft/mission:** STS-129
U.S. EVA 223	**Crew:** Charles Hobaugh, Barry Wilmore, Michael Foreman, Leland Melvin, Robert Satcher, Nicole Stott (returned to Earth), Randolph Bresnik
Shuttle EVA 144	**Spacewalkers:** Robert Satcher, Randolph Bresnik
	Purpose: Third of three scheduled EVAs. Install O_2 tank outside Quest airlock; install MISSE 7 experiment; install common attach system on S3 truss; install thermal covers on MBS; install two fluid jumpers

20. "AIS on ISS: Assembling the Experiment," accessed 2 June 2014, European Space Agency, (*http://www.esa.int/Our_Activities/Technology/AIS_on_ISS_Assembling_the_experiment*); "STS-129 Flight Day 6 Gallery," Amiko Kauderer, editor, accessed 2 June 2014, (*http://www.nasa.gov/mission_pages/shuttle/shuttlemissions/sts129/multimedia/fd6/fd6_gallery.html*); "STS-129: Stocking the Station," National Aeronautics and Space Administration, accessed 2 June 2014, (*http://www.nasa.gov/pdf/398418main_sts129_press_kit.pdf*).

21. Ibid.

A drinking-water valve in Satcher's spacesuit dislodged and postponed the last EVA of 2009 for more than an hour. After they reopened his helmet and reaffixed the valve, the spacewalk began. Satcher and Bresnik immediately retrieved the high pressure O_2 tank from Atlantis's cargo bay and installed it outside Quest. The tank was later used to pressurize and depressurize the airlock. Bresnik then installed the MISSE 7 experiment on the Expedite the Processing of Experiments to the Space Station (ExPRESS) Logistics Carrier, which is a truss-mounted pallet that attaches experiments to the Station's exterior. At the end of the excursion, the spacewalkers deployed another CAS on the S3 truss. Though they had a late start, the spacewalk ended nearly on time.[22]

27 November STS-129/Atlantis landing

1 December Soyuz-TMA 15/ISS Expeditions 20 and 21 landing

20 December Soyuz-TMA 17/ISS Expeditions 22 and 23 launch

22. Ibid.

2010

14 JANUARY

2010 EVA 1	**Duration:** 5:44
World EVA 326	**Spacecraft/mission:** International Space Station Expedition 22
Russian EVA 131	**Crew:** Jeffrey Williams (NASA), Timothy Creamer (NASA), Oleg Kotov (Russian Space Agency), Soichi Noguchi (JAXA), Maxim Suraev (Russian Space Agency)
Space Station EVA 151	**Spacewalkers:** Maxim Suraev, Oleg Kotov
ISS EVA 46	**Purpose:** Prep MRM-2 for Soyuz/Progress usage by installing Kurs navigation antennas and cables and docking targets; install two hatch handrails; jettison excess thermal blankets; retrieve Biorisk experiment

This was the only spacewalk carried out during ISS Expedition 22. Maxim Suraev and Kotov performed work on MRM-2, which docked to Zvezda in November 2009. Their work was similar to the work that Barratt and Padalka did on Zvezda in preparation for the arrival of MRM-2. The cosmonauts installed Kurs navigation antennas, cables, and a docking target on the new module. They also installed handrails to assist in EVA work. After finishing their work on MRM-2, Suraev and Kotov retrieved the Biorisk experiment canister from outside the Station and brought it inside, concluding their spacewalk.[1]

8 February STS-130/Endeavour launch

11 FEBRUARY

2010 EVA 2	**Duration:** 6:32
World EVA 327	**Spacecraft/mission:** STS-130
U.S. EVA 224	**Crew:** George Zamka, Terry Virts, Robert Behnken, Kathryn Hire, Nicholas Patrick, Stephen Robinson
Shuttle EVA 145	**Spacewalkers:** Robert Behnken, Nicholas Patrick
	Purpose: First of three scheduled EVAs. Remove covers and heater cables from stowed Tranquility node; install Tranquility to Unity; prepare temporary platform in worksite; connect Tranquility to avionics and heater cables; release Tranquility gap spanners

1. "Expedition 19 and 20: Full Partners," National Aeronautics and Space Administration, accessed 21 June 2014, (*http://www.nasa.gov/pdf/320539main_Expedition_19_20_Press_Kit.pdf*); "Station Crew Completes First Expedition 22 Spacewalk, Amiko Kauderer, editor, accessed 3 June 2014, (*http://www.nasa.gov/mission_pages/station/expeditions/expedition22/exp22_eva.html*).

FIGURE 9. **Cupola.** NASA astronaut Robert Behnken, STS-130 mission specialist, poses for a photo near the windows in the ISS Cupola while Space Shuttle Endeavour remains docked with the Station. (NASA S130-E-010477)

The STS-130 mission delivered the remaining two U.S.-made components to the ISS: the Tranquility node and the 7-windowed Cupola control room. Tranquility, one of the final modules added to the ISS, provides additional space for life support and environmental control systems. It also provides another docking port for supply ships and Shuttles. The Cupola is a work module connected to Tranquility. It is known for its six expansive side windows and a wide ceiling window. Behnken first moved to the worksite and prepared Unity for its new attachment. He readied a camera on the 10-year-old node, which offered another view during docking. Behnken also detached eight contamination covers that protected the berthing port. Meanwhile, Patrick removed Tranquility's protective covers and detached its insulation cables from Endeavour. Next, from within the Station, astronauts Terry Virts and Kathryn Hire robotically transferred Tranquility out of the payload bay and docked it to Unity. The spacewalkers then prepared Dextre and the required tools for cable installations. Behnken and Patrick linked the new node to Unity's heater lines and eight avionics cables. During the next EVA, the pair further integrated Tranquility to the Station.[2]

2. "Space Station's Tranquility," Cheryl Mansfield, accessed 12 June 2014, (*http://www.nasa.gov/centers/kennedy/stationpayloads/tranquility.html*); "Station Gain Unparalleled Views," Cheryl Mansfield, accessed 12 June 2014, (*http://www.nasa.gov/mission_pages/shuttle/shuttlemissions/sts130/launch/130_overview.html*); "STS-130: A Room with a View," National Aeronautics and Space Administration, accessed 12 June 2014, (*http://www.nasa.gov/pdf/423514main_sts130_press_kit_2.pdf*).

13 FEBRUARY

2010 EVA 3	**Duration:** 5:54
World EVA 328	**Spacecraft/mission:** STS-130
U.S. EVA 225	**Crew:** George Zamka, Terry Virts, Robert Behnken, Kathryn Hire, Nicholas Patrick, Stephen Robinson
Shuttle EVA 146	**Spacewalkers:** Robert Behnken, Nicholas Patrick
	Purpose: Second of three scheduled EVAs. Install ammonia lines from Destiny to Tranquility; install heater cables and thermal covers on Tranquility; install negative pressure vent, eight handrails, and gap spanners; install centerline camera by berthing port

Behnken and Patrick continued assembling Tranquility and prepared the Cupola for installation. During the 4 initial hours, the pair coupled two ammonia loops between Tranquility and the Destiny laboratory. Each loop, comprised of two fluid lines, was strung from Destiny, through brackets on Unity, and finally to Tranquility. The spacewalkers then focused on assembling the new node's external components. Behnken mounted an insulation cover over Tranquility keel pins and four trunnions. He also assembled a centerline camera and removed launch locks on Tranquility, to allow for the Cupola attachment during the next excursion. Meanwhile, Patrick installed eight handrails and the exhaust valve of an atmospheric control and resupply system.[3]

16 FEBRUARY

2010 EVA 4	**Duration:** 5:48
World EVA 329	**Spacecraft/mission:** STS-130
U.S. EVA 226	**Crew:** George Zamka, Terry Virts, Robert Behnken, Kathryn Hire, Nicholas Patrick, Stephen Robinson
Shuttle EVA 147	**Spacewalkers:** Nicholas Patrick, Robert Behnken
	Purpose: Third of three scheduled EVAs. Connect heater and data cables to PMA-3; open ammonia loop 3 to Tranquility; remove Cupola thermal covers and release launch locks on windows; install five handrails and foot restraints on Tranquility; route video signal converter; cover Harmony node's centerline camera

Prior to the excursion, Hire and Virts robotically transferred the PMA-3 and the Cupola to their final home on Tranquility. During the last EVA, Behnken and Patrick completed exterior assemblage on the node and its Cupola. First, they ran ammonia through the second fluid line between Destiny and the new node. Behnken continued installing five more handrails and foot restraints on Tranquility. Meanwhile, Patrick connected heater and data cables between PMA-3 and the new node. He also

---

3. "Station Gains Unparalleled Views," Cheryl Mansfield, accessed 12 June 2014, (*http://www.nasa.gov/mission_pages/shuttle/shuttlemissions/sts130/launch/130_overview.html*); "STS-130: A Room with a View," National Aeronautics and Space Administration, accessed 12 June 2014, (*http://www.nasa.gov/pdf/423514main_sts130_press_kit_2.pdf*).

removed the six insulation covers on the Cupola windows and the launch locks securing the window shutters. For the first time, the windows were open and exposed to the space environment. Finally, the spacewalkers routed video signal converter cables from the complex of cable connections on the S0 truss to the Zarya module. This allowed operation of the Station's Canadarm2 from the Russian portion of the Station.[4]

21 February STS-130/Endeavour landing

18 March Soyuz-TMA 16/ISS Expeditions 21 and 22 landing

2 April Soyuz-TMA 18/ISS Expeditions 23 and 24 launch

5 April STS-131/Discovery launch

9 APRIL

2010 EVA 5
World EVA 330
U.S. EVA 227
Shuttle EVA 148

Duration: 6:27

Spacecraft/mission: STS-131

Crew: Alan Poindexter, James Dutton, Clayton Anderson, Rick Mastracchio, Dorothy Metcalf-Lindenburger, Stephanie Wilson, Naoko Yamazaki (JAXA)

Spacewalkers: Rick Mastracchio, Clayton Anderson

Purpose: First of three scheduled EVAs. Temporarily stow new ammonia tank on robotic arm; retrieve MPAC-SEED experiment from Kibo; replace rate gyro assembly; prepare for battery replacement

The main objective of the three STS-131 mission spacewalks was to assemble a new ammonia tank and store the depleted tank in Discovery's cargo bay. Though seemingly simple, it required a complex sequence of unloading and restoring both tanks and changing the Canadarm2 base twice. Anderson and Mastracchio also found some time between their primary work to complete get-ahead tasks. At the start of the excursion, Mastracchio prepared the new tank for assembly while Anderson began detaching the exhausted tank. Mastracchio first prepared the new ammonia tank in Discovery's cargo bay. He unbolted the tank's four launch locks and then mounted a handle on the tank, which the robotic arm later clutched. Meanwhile, Anderson uncoupled the depleted tank's four ammonia and nitrogen lines. They then met at the cargo bay and positioned the new tank to be grasped by the robotic arm and driven to the Quest airlock. They again separated and took on different work. Anderson cleaned up the worksite and Mastracchio retrieved the MPAC/SEED for return to Earth. The pair then installed a second handle on the new ammonia tank, which the robotic arm grasped to pull to a temporary stowage area. As the tank transferred, the spacewalkers took on the get-ahead tasks. They replaced the rate gyro assembly and readied P6 truss batteries for replacement.[5]

---

4. Ibid.

5. "Mission Archives: STS-131," Amiko Kauderer, editor, accessed 8 March 2014, (*http://www.nasa.gov/mission_pages/shuttle/shuttlemissions/sts131/main/index.html*); "STS-131: Experiment Express," National Aeronautics and Space Administration, accessed 13 June 2014, (*http://www.nasa.gov/pdf/440897main_sts_131_press_kit.pdf*).

11 APRIL

2010 EVA 6	**Duration:** 7:26
World EVA 331	**Spacecraft/mission:** STS-131
U.S. EVA 228	**Crew:** Alan Poindexter, James Dutton, Clayton Anderson, Rick Mastracchio, Dorothy Metcalf-Lindenburger, Stephanie Wilson, Naoko Yamazaki (JAXA)
Shuttle EVA 149	**Spacewalkers:** Rick Mastracchio, Clayton Anderson
	Purpose: Second of three scheduled EVAs. Remove old ammonia tank and temporarily stow on CETA cart; assemble new ammonia tank; install two radiator grapple beams; retrieve debris shields in external stowage platform

Anderson and Mastracchio focused on storing the exhausted tank for the majority of the excursion. They disconnected its two power cables and four bolts, and then handed the tank to the robotic arm. Next, it was transferred and secured to the CETA cart with six tethers for temporary storage. While they waited for the arm to seize the new ammonia tank, Anderson and Mastracchio installed two radiator grapple stowage beams on the P1 truss. Future missions utilized the beams to store spare equipment. Next, the spacewalkers returned to the S1 truss and successfully secured the new tank with four bolts, though they had trouble screwing in one. They then connected the tank to its six cables. Anderson and Mastracchio untied the old tank from the CETA cart, permitting the robotic arm to stow it on the Mobile Transporter temporarily. Their last task before finalizing the excursion was returning two debris shields from the external stowage platform 2 to Quest.[6]

13 APRIL

2010 EVA 7	**Duration:** 6:24
World EVA 332	**Spacecraft/mission:** STS-131
U.S. EVA 229	**Crew:** Alan Poindexter, James Dutton, Clayton Anderson, Rick Mastracchio, Dorothy Metcalf-Lindenburger, Stephanie Wilson, Naoko Yamazaki (JAXA)
Shuttle EVA 150	**Spacewalkers:** Rick Mastracchio, Clayton Anderson
	Purpose: Third of three scheduled EVAs. Complete new ammonia tank installation; stow old ammonia tank in Discovery's payload bay

Prior to the spacewalk, the crew remotely operated the robotic arm to retrieve the spent tank from the mobile transporter. Anderson and Mastracchio first finalized the installation of the new tank, and then worked on the spent tank in Discovery's payload bay and secured it with four bolts. Approximately an hour into the walk, they completed their main objective. The pair dedicated the remainder of the excursion to get-ahead tasks. Anderson rode the robotic arm to the Columbus laboratory and stowed a lightweight adapter plate assembly in the payload bay. The device stored experiments on Columbus's

---

6. "Mission Archives: STS-131," Amiko Kauderer, editor, accessed 8 March 2014 (*http://www.nasa.gov/mission_pages/shuttle/shuttlemissions/sts131/main/index.html*); "STS-131: Experiment Express," National Aeronautics and Space Administration, accessed 13 June 2014, (*http://www.nasa.gov/pdf/440897main_sts_131_press_kit.pdf*).

exterior. Next, Anderson installed a second video camera and removed a redundant insulation blanket on Dextre. In the meantime, Mastracchio replaced a failed light on Destiny, and then installed two radiator grapple fixture stowage beams on the Station's starboard side. Finally, he installed the Worksite Interface Extender ORU on the ISS robotic arm Mobile Transporter.[7]

20 April STS-131/Discovery landing

14 May STS-132/Atlantis launch

17 MAY

2010 EVA 8

World EVA 333

U.S. EVA 230

Shuttle EVA 151

Duration: 7:25

Spacecraft/mission: STS-132

Crew: Kenneth Ham, Dominic Antonelli, Michael Good, Garrett Reisman, Piers Sellers, Stephen Bowen

Spacewalkers: Stephen Bowen, Garrett Reisman

Purpose: First of three scheduled EVAs. Install spare Ku antenna boom and dish on Z1 truss; install storage platform for Dextre; loosen P6 battery bolts

The main objective of the STS-132 mission was to ensure all Station components were in outstanding condition and well enough to endure far into the future. With the end of the Space Shuttle Program in sight, NASA wanted to put the final touches on the Station. In Atlantis's payload bay, the crew delivered various spare components and Rassvet, a Russian laboratory module also known as the Mini-Research Module-1 (MRM-1). The almost 20-foot (6.09-meter) long and 17,700-pound (7.71-kilogram) laboratory is a relatively small module. The spacewalkers' priority was the battery replacement, though they also found time to complete other tasks. Once they delivered tools and spare parts to the worksite, Bowen and Reisman turned their attention to first installing the spare Ku antenna boom. They then attached the 6-foot (1.83-meter) diameter high-speed communications dish antenna to the boom. The antenna provides two-way data and video connections between ground controllers and Station residents. Bowen mounted the standby 6-foot (1.83-meter) Ku antenna boom with two bolts onto the Z1 truss. He then connected its six power and data cables. Above Destiny, the pair then assembled a new storage platform for Dextre. With time remaining, the spacewalkers completed a get-ahead task for the two following EVAs. Bowen and Reisman loosened the 12 bolts holding the P6 truss's batteries and concluded the walk.[8]

7. "Final Planned Flight of *Atlantis* Delivers New 'Dawn'," Anna Heiney, accessed 14 June 2014, (*http://www.nasa.gov/mission_pages/shuttle/shuttlemissions/sts132/launch/132_overview.html*); "STS-132: Finishing Tools," National Aeronautics and Space Administration, accessed 14 June 2014, (*http://www.nasa.gov/pdf/451029main_sts132_press_kit2.pdf*).

8. "Final Planned Flight of Atlantis Delivers New 'Dawn'," Anna Heiney, accessed 14 June 2014, (*http://www.nasa.gov/mission_pages/shuttle/shuttlemissions/sts132/launch/132_overview.html*); "STS-132: Finishing Tools," National Aeronautics and Space Administration, accessed 14 June 2014, (*http://www.nasa.gov/pdf/451029main_sts132_press_kit2.pdf*).

19 MAY

2010 EVA 9	**Duration:** 7:09
World EVA 334	**Spacecraft/mission:** STS-132
U.S. EVA 231	**Crew:** Kenneth Ham, Dominic Antonelli, Michael Good, Garrett Reisman, Piers Sellers, Stephen Bowen
Shuttle EVA 152	**Spacewalkers:** Stephen Bowen, Michael Good
	Purpose: Second of three scheduled EVAs. Unsnag cable of orbiter inspection boom sensor; replace four batteries on P6 truss; bolt Ku antenna dish to its boom; release locks on dish and allow it to rotate

The second excursion of the mission was designed to be entirely devoted to replacing old batteries on the P6 truss; however, a snagged cable connected to the orbiter inspection boom sensor delayed the main objective. In addition to unsnagging the cable, Bowen and Good bolted the Ku antenna dish to its boom and released the locks restricting its rotation. Even with the delay, the spacewalkers replaced four batteries, one more battery than scheduled. The batteries are on the truss's B side arrays. The truss has two sets of solar arrays, known as the "A side" and the "B side." The Side A batteries were replaced during the STS-127 mission. Due to the need for careful removal, handling, transfer, and temporary stowage of the failed batteries and then the handling and replacement of the new batteries, the crew referred to these EVA operations as the "shepherding technique." Various EVA tools and equipment were used to remove the batteries, including the EVA Power Tool, ORU handling tools, and the Integrated Cargo Carrier-Vertical Light Deployable (ICC-VLD) for storing the batteries.[9]

21 MAY

2010 EVA 10	**Duration:** 6:46
World EVA 335	**Spacecraft/mission:** STS-132
U.S. EVA 232	**Crew:** Kenneth Ham, Dominic Antonelli, Michael Good, Garrett Reisman, Piers Sellers, Stephen Bowen
Shuttle EVA 153	**Spacewalkers:** Michael Good, Garrett Reisman
	Purpose: Third of three scheduled EVAs. Install nitrogen jumper port trusses; replace two batteries on P6 truss

Reisman replaced Bowen and worked with Good to finalize the battery replacement on the P6 truss. The spacewalkers used the same method as in the previous spacewalk to complete the work. Additionally, Good and Reisman had time to install a reserve nitrogen jumper between the P4 and P5 trusses, in case of a break in the Station's coolant system.[10]

9. Ibid.

10. Ibid.

FIGURE 10. **Rassvet.** The Soyuz TMA-19 spacecraft (foreground), docked to the Rassvet Mini-Research Module 1 (MRM1), and Progress 37 resupply vehicle, docked to the Pirs Docking Compartment, are featured in this image photographed by an Expedition 24 crewmember on the ISS. (NASA ISS024-E-007122)

26 May	STS-132/Atlantis landing
1 June	Soyuz-TMA 17/ISS Expeditions 22 and 23 landing
15 June	Soyuz-TMA 19/ISS Expeditions 24 and 25 launch

27 JULY

2010 EVA 11
World EVA 336
Russian EVA 132
Space Station EVA 152
ISS EVA 47

Duration: 6:42

Spacecraft/mission: International Space Station Expedition 24

Crew: Alexander Skvortsov (Russian Space Agency), Tracy Caldwell Dyson (NASA), Mikhail Kornienko (Russian Space Agency), Shannon Walker (NASA), Douglas Wheelock (NASA), Fyodor Yurchikhin (Russian Space Agency)

Spacewalkers: Fyodor Yurchikhin, Mikhail Kornienko

Purpose: Replace and jettison old ATV camera; install new ATV camera; attach cables to Mini-Research Module 1

Orlan-clad cosmonauts Fyodor Kornienko and Yurchikhin egressed the airlock located in the Pirs Docking Compartment module to begin installing the Mini-Research Module 1 (MRM-1) cables. The Russian-made module, more commonly known as "Rassvet," has a docking port for Russian vehicles. The pair also routed and powered the Command and Data Handling cables needed to operate the Zarya and Zvezda modules. Additionally, the spacewalkers jettisoned a degraded ATV camera and replaced it with a more advanced camera. The new device provides video images of future docking ATVs.[11]

7 AUGUST

2010 EVA 12

World EVA 337

U.S. EVA 233

Space Station EVA 153

ISS EVA 48

Duration: 8:03

Spacecraft/mission: International Space Station Expedition 24

Crew: Alexander Skvortsov (Russian Space Agency), Tracy Caldwell Dyson (NASA), Mikhail Kornienko (Russian Space Agency), Shannon Walker (NASA), Douglas Wheelock (NASA), Fyodor Yurchikhin (Russian Space Agency)

Spacewalkers: Douglas Wheelock, Tracy Caldwell Dyson

Purpose: Detach four nitrogen connectors from failed ammonia pump

The first of three spacewalks to replace the failed coolant pump on the S1 truss required over 8 hours, making it the longest ISS spacewalk and the sixth longest when including Shuttle EVAs. Wheelock and first-time spacewalker Tracy Caldwell Dyson first attempted to disconnect the four ammonia lines from the failed pump. Three cooling lines were successfully removed; however, the final connector leaked and contaminated the spacesuits. They decided to reattach the connection and use a spool-positioning device to ensure correct pressure within the ammonia route. Caldwell Dyson and Wheelock ended the walk and decontaminated their suits.[12]

11 AUGUST

2010 EVA 13

World EVA 338

U.S. EVA 234

Space Station EVA 154

ISS EVA 49

Duration: 7:26

Spacecraft/mission: International Space Station Expedition 24

Crew: Alexander Skvortsov (Russian Space Agency), Tracy Caldwell Dyson (NASA), Mikhail Kornienko (Russian Space Agency), Shannon Walker (NASA), Douglas Wheelock (NASA), Fyodor Yurchikhin (Russian Space Agency)

Spacewalkers: Douglas Wheelock, Tracy Caldwell Dyson

Purpose: Remove fourth ammonia connector; detach old ammonia pump electrical cables; unbolt old ammonia pump and stow in MBS; disconnect electrical cables to new ammonia pump

---

11. "Cosmonauts Complete First Expedition 24 Spacewalk," Amiko Kauderer, editor, accessed 20 March 2014, (*http://www.nasa.gov/mission_pages/station/expeditions/expedition24/russian_eva25.html*).

12. "Expedition 24 Performs First Spacewalk to Replace Ammonia Pump," Amiko Kauderer, editor, accessed 21 March 2014, (*http://www.nasa.gov/mission_pages/station/main/080710_spacewalk.html*).

Caldwell Dyson and Wheelock were more successful during their second excursion to swap the failed ammonia coolant pump. This time, Wheelock closed the ammonia valve before he disconnected the leaking line. Caldwell Dyson then detached five electrical and data cables from the old pump, while Wheelock unfastened its four bolts. They then stowed the pump on the MBS payload bracket. Before they ended the excursion, the pair disconnected the new pump's power cables in preparation for the following spacewalk.[13]

16 AUGUST

2010 EVA 14
World EVA 339
U.S. EVA 235
Space Station EVA 155
ISS EVA 50

Duration: 7:20

Spacecraft/mission: International Space Station Expedition 24

Crew: Alexander Skvortsov (Russian Space Agency), Tracy Caldwell Dyson (NASA), Mikhail Kornienko (Russian Space Agency), Shannon Walker (NASA), Douglas Wheelock (NASA), Fyodor Yurchikhin (Russian Space Agency)

Spacewalkers: Douglas Wheelock, Tracy Caldwell Dyson

Purpose: Install new coolant pump; stow support equipment

With the assistance of Shannon Walker within the Station, Caldwell Dyson and Wheelock finalized the installation of the spare 780-pound (353.80-kilogram) ammonia pump on the S1 truss. Walker operated the Canadarm2 and drove the module to its new home. Wheelock then secured its four bolts and Caldwell Dyson coupled its five electrical cables. After ground controllers confirmed its pressure and connections, the spacewalkers opened the ammonia valve and filled the pump.[14]

25 September Soyuz-TMA 18/ISS Expeditions 23 and 24 landing

7 October Soyuz-TMA 01M/ISS Expeditions 25 and 26 launch

15 NOVEMBER

2010 EVA 15
World EVA 340
Russian EVA 133
Space Station EVA 156
ISS EVA 51

Duration: 6:28

Spacecraft/mission: International Space Station Expedition 25

Crew: Douglas Wheelock (NASA), Shannon Walker (NASA), Alexander Kaleri (Russian Space Agency), Scott Kelly (NASA), Oleg Skripochka (Russian Space Agency), Fyodor Yurchikhin (Russian Space Agency)

Spacewalkers: Fyodor Yurchikin, Oleg Skripochka

Purpose: Install multipurpose workstation on Zvezda; clean Kontur experiment on Zvezda; install experiments on Rassvet; reposition camera on Rassvet

13. "Expedition 24 Performs Second Spacewalk to Replace Ammonia Pump," Amiko Kauderer, editor, accessed 21 March 2014, (*http://www.nasa.gov/mission_pages/station/expeditions/expedition24/081110_spacewalk.html*).

14. "Spacewalkers Install Space Ammonia Pump," National Aeronautics and Space Administration, accessed 21 March 2014, (*http://www.nasa.gov/mission_pages/station/expeditions/expedition24/081610_spacewalk.html*).

Skripochka and Yurchikhin left from Pirs to execute the one and only spacewalk during Expedition 25. The two cosmonauts installed a multipurpose workstation on the large diameter section on Zvezda's starboard side. Upon completion of this task, the duo moved on to research collection and maintenance tasks. They started by cleaning and moving the Kontur (short for Development of a System of Supervisory Control Over the Internet of the Robotic Manipulator in the Russian Segment of ISS) robotics experiment from the port side of Zvezda into the Pirs airlock. They then attached a new materials experiment to a handrail on the Rassvet module. Skripochka and Yurchikhin temporarily stored samples beneath the insulation outside of Zvezda and Pirs to bring back inside the Station for analysis later. The cosmonauts were also tasked with moving a television camera from one end of the Rassvet's docking compartment to another, but the move was unsuccessful due to the camera interfering with the insulation in the new location.[15]

25 November Soyuz-TMA 19/ISS Expeditions 24 and 25 landing

15 December Soyuz-TMA 20/ISS Expeditions 26 and 27 launch

15. "Cosmonauts Perform 26th Russian Space Station Spacewalk," Amiko Kauderer, editor, accessed 22 March 2014, (*http://www.nasa.gov/mission_pages/station/expeditions/expedition25/russian_eva26.html*).

2011

21 JANUARY

2011 EVA 1	**Duration:** 5:23
World EVA 341	**Spacecraft/mission:** International Space Station Expedition 26
Russian EVA 134	**Crew:** Scott Kelly (NASA), Catherine Coleman (NASA), Alexander Kaleri (Russian Space Agency), Dmitry Kondratyev (Russian Space Agency), Paolo Nespoli (European Space Agency), Oleg Skripochka (Russian Space Agency)
Space Station EVA 157	
ISS EVA 52	**Spacewalkers:** Dmitry Kondratyev, Oleg Skripochka
	Purpose: Install 100Mb-per-second data downlink antenna and its cables; retrieve plasma and Expose-R experiments; finish installation of MRM-1 docking TV camera; jettison antenna cover and cable reel

When Dmitry Kondratyev and Skripochka set out on the first EVA of Expedition 26, NASA equipped Skripochka's helmet with lights and a wireless television camera to provide a live point-of-view feed to mission controls in Houston and Moscow. As they were tasked with several jobs during this EVA, they clipped all of their tool carriers and new equipment to Zvezda's exterior for easy access and storage between the jobs. Firstly, they deployed an antenna to be used for the radio technical system for information transfer. The experimental system was designed to allow NASA, RSA, ESA, and other space agencies on Earth to downlink data files from ISS at the speed of about 100 Mb per second using radio signals. Once they erected the antenna, they routed external cables, connecting it to patch panels that then connected it to computers within ISS. Upon completion of this task, the cosmonauts jettisoned the antenna's hatbox-shaped cover and its cable reel. Kondratyev and Skripochka then retrieved two experiments hanging on ISS's exterior. One was a plasma pulse generator that was supposed to measure disturbances in the ionosphere caused by the Station's plasma impulse flow, but the test experienced an early fail. The other was the Expose-R experiment canister. The spacewalkers retrieved both from the port side of Zvezda and placed them in Pirs, along with a tool bag they used while installing the antenna. After placing the experiments in the airlock, they grabbed a docking camera left to install on Rassvet. Skripochka and Yurchikhin had tried to install this camera on 15 November 2010, but they encountered problems getting past the insulation. For this spacewalk, Skripochka and Kondratyev equipped themselves with special cutters to tear through the insulation threading, which allowed them to work around the camera mount unobstructed. After installing the camera, they connected its cable to a prewired connector, which fed the video to the Station, and they returned to Pirs airlock.[1]

1. "Cosmonauts Perform 27th Russian Space Station Spacewalk," Amiko Kauderer, editor, accessed 23 March 2014, (*http://www.nasa.gov/mission_pages/station/expeditions/expedition26/russian_eva27.html*).

16 FEBRUARY

2011 EVA 2	**Duration:** 4:51
World EVA 342	**Spacecraft/mission:** International Space Station Expedition 26
Russian EVA 135	**Crew:** Scott Kelly (NASA), Catherine Coleman (NASA), Alexander Kaleri (Russian Space Agency), Dmitry Kondratyev (Russian Space Agency), Paolo Nespoli (European Space Agency), Oleg Skripochka (Russian Space Agency)
Space Station EVA 158	**Spacewalkers:** Dmitry Kondratyev, Oleg Skripochka
ISS EVA 53	**Purpose:** Install Molniya-Gamma and Radiometria experiments on SM; jettison foot restraint; retrieve Komplast panels from FGB

Kondratyev and Skripochka once again joined forces for an EVA. Their Orlan space suits were both fitted again with the NASA-provided helmet lights and television cameras that provide live video to both the Moscow and Houston Mission Control Centers. After removing the two experiments from Zvezda in the previous spacewalk, they installed two new ones. One was Radiometria, which would gather useful data for seismic forecasts and the prediction of earthquakes. The other experiment was Molniya-Gamma, which was to study gamma splashes and optical radiation during terrestrial lighting and thunderstorms. They installed both experiments atop the portable workstations, with Radiometria being located on Zvezda's port side, and Molniya-Gamma ending up on starboard side. From the Zarya module, they retrieved the Komplast experiment panels, which, after having been exposed to space, would be used to determine the best materials to use in building a long-duration spacecraft. Then, before returning to Pirs, they detached and jettisoned a foot restraint.[2]

24 February STS-133/Discovery launch

28 FEBRUARY

2011 EVA 3	**Duration:** 6:34
World EVA 343	**Spacecraft/mission:** STS-133
U.S. EVA 236	**Crew:** Steven Lindsey, Eric Boe, Michael Barratt, Stephen Bowen, Alvin Drew, Nicole Stott
Shuttle EVA 159	**Spacewalkers:** Stephen Bowen, Alvin Drew
ISS EVA 54	**Purpose:** First of two scheduled EVAs. Install backup power cable for new stowage module; relocate failed pump module and prep it for NH_3 venting; reposition Z1 truss RPCM thermal blanket and EVA foot restraint socket extension; retrieve foot restraint for in-cabin maintenance; add wedge to S1 camera; enable MT and CETA cart translation past starboard solar array joint by installing rail extensions and removing stops; collect vacuum in JAXA experiment bottle

2. "Cosmonauts Perform 28th Russian Space Station Spacewalk," Amiko Kauderer, editor, accessed 24 March 2014, (*http://www.nasa.gov/mission_pages/station/expeditions/expedition26/russian_eva28.html*).

This was Alvin Drew's first EVA, and he became the 200th person to walk in space. He and Bowen began by installing an extension cable from the Unity to the Tranquility Node. With Drew and Bowen looking on, Scott Kelly and Barratt moved the ISS's failed 800-pound (362.88-kilogram) ammonia pump module from Canadarm2's mobile base to an external stowage platform. There was initially some trouble completing this task because Barratt and station commander Scott Kelly were unable to control Canadarm2 from the Cupola, but they moved to another control station for the arm in Destiny and had no further problems. Drew then retrieved a tool and stowed it with the module to use in draining the remaining ammonia on the next spacewalk. Meanwhile, Bowen installed a foot restraint on the arm of ISS. Both astronauts removed tethers and cart stoppers along the Mobile Transporter railway, then added extensions to the track's rails. Drew and Bowen also installed a camera wedge on the Station. They concluded the spacewalk by working with Japan's "Message in a Bottle," a canister autographed by different astronauts. They opened the canister to gather some of the vacuum of space, sealed it, and then returned it to Discovery for return to Earth for public display.[3]

2 MARCH

2011 EVA 4 **World EVA 344** **U.S. EVA 237** **Shuttle EVA 155**	**Duration:** 6:14 **Spacecraft/mission:** STS-133 **Crew:** Steven Lindsey, Eric Boe, Michael Barratt, Stephen Bowen, Alvin Drew, Nicole Stott **Spacewalkers:** Stephen Bowen, Alvin Drew **Purpose:** Second of two scheduled EVAs. Install light on port CETA cart, camera on SPDM robot, and 3 camera lens covers; reposition camera sunshade for clear view; resecure loose radiator grapple beam; retrieve pallet with MISSE data; remove thermal insulation from several locations; reinstall foot restraint and relocate another; relocate Strela grapple fixture adapter to FGB

This EVA was delayed by 24 minutes due to the crew replacing an O-ring on the lithium hydroxide canister of Bowen's spacesuit to fix a small leak. The two astronauts worked separately during most of the spacewalk. Bowen briefly set up Canadarm2, and then, from the Columbus module, retrieved a lightweight adapter plate assembly, which would hold experiments, and installed it in Discovery's cargo bay. Drew meanwhile retrieved the tool that he left with the ammonia pump module on the previous EVA and used it to vent the remaining coolant left in the module. He then returned the tool to the airlock. The astronauts reunited near Dextre, upon which Bowen installed a tilt assembly and light pan, and removed some insulation. Drew also removed some insulation and jettisoned it. He repositioned a sunshade on a camera near Dextre. He then moved on to install a light on one of the handcars on one of the truss's rails. Drew then returned to Canadarm2 to put a lens cover over the camera on its elbow. He also repositioned a foot restraint and a Russian cargo arm adapter on the arm. Drew then continued by making repairs on the insulation on a radiator beam valve module on the port truss. He

3. "STATUS REPORT: STS-133-09," Amiko Kauderer, editor, accessed 24 March 2014, (*http://www.nasa.gov/mission_pages/shuttle/shuttlemissions/sts133/news/STS-133-09.html*).

then secured a stowage beam grapple fixture. When nearing the end of their spacewalk, the light atop Drew's helmet fell off his suit. Bowen tried to reattach it, but was unsuccessful, and they attached it to a tether to bring it back to the airlock with them as they returned.[4]

9 March	STS-133/Discovery landing
16 March	Soyuz-TMA 01M/ISS Expeditions 25 and 26 landing
4 April	Soyuz-TMA 21/ISS Expeditions 27 and 28 launch
16 May	STS-134/Endeavour launch

20 MAY

2011 EVA 5

World EVA 345

U.S. EVA 238

Shuttle EVA 156

Duration: 6:19

Spacecraft/mission: STS-134

Crew: Mark Kelly, Gregory Johnson, Gregory Chamitoff, Andrew Feustel, Michael Fincke, Roberto Vittori (European Space Agency)

Spacewalkers: Andrew Feustel, Gregory Chamitoff

Purpose: First of four scheduled EVAs. Retrieve MISSE 7 experiment; install MISSE 8; install CETA cart light; reinstall SARJ cover; install ammonia jumper; vent nitrogen lines; install External Wireless Communication System on Destiny

Endeavour delivered the Alpha Magnetic Spectrometer-2 (AMS-02), two spare S-band antennas, a high-pressure gas tank, and spare components for Dextre. During the mission's first of four EVAs, Chamitoff and Feustel replaced the MISSE experiments and installed the CETA cart light and the ammonia jumper located on the S3 truss. Feustel replaced the MISSE 7 experiments with the new MISSE 8. The former returned to Earth in Endeavour's payload bay. After Chamitoff bolted and powered the new CETA cart light, the spacewalkers made the necessary preparations for the following excursion to refill the leaking ammonia pump. Chamitoff and Feustel installed a new ammonia jumper and vented the nitrogen between the P1 and P5 trusses. Next, the duo attempted to install the External Wireless Communication System on the Destiny laboratory. While installing the two wireless antennas, Chamitoff received a "CO_2 SENSOR BAD" message on his EMU Caution and Warning System. This led to the EVA being cut short.[5]

---

4. "STATUS REPORT: STS-133-13," Amiko Kauderer, editor, accessed 28 March 2014, (*http://www.nasa.gov/mission_pages/shuttle/shuttlemissions/sts133/news/STS-133-13.html*).

5. "STS-134: Final Flight of *Endeavour*," National Aeronautics and Space Administration, accessed 25 June 2014, (*http://www.nasa.gov/pdf/538352main_sts134_presskit_508.pdf*); "STS-134 MISSION," Jeanne Ryba, editor, accessed 25 June 2014, (*http://www.nasa.gov/mission_pages/shuttle/shuttlemissions/sts134/launch/index.html*).

22 MAY

2011 EVA 6	**Duration:** 8:07
World EVA 346	**Spacecraft/mission:** STS-134
U.S. EVA 239	**Crew:** Mark Kelly, Gregory Johnson, Gregory Chamitoff, Andrew Feustel, Michael Fincke, Roberto Vittori (European Space Agency)
Shuttle EVA 157	**Spacewalkers:** Andrew Feustel, Michael Fincke
	Purpose: Second of four scheduled EVAs. Reconnect ammonia loops and refill ammonia tank; vent refilled lines; lubricate SARJ; install cover over Dextre camera; lubricate Dextre wrist joints; install two radiator grapple bar stowage beams on S1 truss

Feustel and Fincke's two primary objectives during the second spacewalk were topping off the ammonia pump tank and lubricating the SARJs. The spacewalkers first continued rerouting the ammonia loops from the previous walk and structured a constant flow of ammonia from the tank on the P1 truss to the leaking loop on the P6. Ground controllers in Houston confirmed the successful reconfiguration. The pair then filled the tank. Before detaching the jumper cable, Feustel vented the residual ammonia. Meanwhile, Fincke climbed to the P3 truss and lubricated the SARJ with a grease gun. Fincke also photographed and obtained samples of the grease from the 2007 STS-126 mission, when the joint was last lubricated. While maneuvering the joint covers, Fincke lost a bolt and three washers. He was forced to return a cover to the Station because the joint had too few bolts. While waiting for the joint to rotate 200 degrees, Fincke had an hour to mount two radiator grapple bar stowage beams on the S1 truss. The beams serve as a storage unit for spacewalk handles. Upon completion, he returned to the SARJ and greased its other side. Meanwhile, Feustel lubricated the Dextre robotic arms and assembled a cover over its cameras. Before ingressing to the airlock, the pair remained in the Sun for 30 minutes in order to evaporate any toxic ammonia crystals from the leakage that may have adhered to their spacesuits.[6]

23 May Soyuz-TMA 20/ISS Expeditions 26 and 27 landing

25 MAY

2011 EVA 7	**Duration:** 6:54
World EVA 347	**Spacecraft/mission:** STS-134
U.S. EVA 240	**Crew:** Mark Kelly, Gregory Johnson, Gregory Chamitoff, Andrew Feustel, Michael Fincke, Roberto Vittori (European Space Agency)
Shuttle EVA 158	**Spacewalkers:** Andrew Feustel, Michael Fincke
	Purpose: Third of four scheduled EVAs. Install Power Data and Grapple Fixture on Zarya; install cables to connect backup power between U.S. and Russian segments; install External Wireless Communication antennas

6. Ibid.

The third excursion centered on amplifying energy routed to the Russian modules and lengthening the reach of the Canadarm2. First, Feustel and Fincke assembled the Power Data and Grapple Fixture, a stand that is bolted to Zarya and allows the robotic arm to reach the module. Still on Zarya, they then installed the video signal converter. Next, the astronauts routed jumper cables from Harmony, to Unity, and finally to Zarya. The cable configuration transfers power between the U.S. and Russian segments. Feustel and Fincke then tackled the unfinished work from the mission's first spacewalk and installed the External Wireless Communication's two wireless antennas. With time remaining, they took on get-ahead tasks for the succeeding EVA. The spacewalkers mounted a thermal blanket on the O_2 tank grapple fixture, photographed the FGB and its thrusters, and secured loose thermal insulation.[7]

27 MAY

2011 EVA 8
World EVA 348
U.S. EVA 241
Shuttle EVA 159

Duration: 7:24

Spacecraft/mission: STS-134

Crew: Mark Kelly, Gregory Johnson, Gregory Chamitoff, Andrew Feustel, Michael Fincke, Roberto Vittori (European Space Agency)

Spacewalkers: Michael Fincke, Gregory Chamitoff

Purpose: Fourth of four scheduled EVAs. Permanently stow orbiter inspection boom on ISS; install Power Data and Grapple Fixture on orbiter inspection boom; release fasteners on spare Dextre arm

In the past, Endeavour's orbiter inspection boom was stowed in the payload bay and returned to Earth with the Shuttle. The boom was a valuable instrument during past EVAs to extend the reach of the Canadarm2, therefore the crew permanently stored it on the Station's truss for future Station maintenance. They then installed a Power Data and Grapple Fixture on the boom to double the arm's previous reach. Previously, the robotic arm could only grasp a fixture in the middle of the boom and therefore only benefitted from half its length. Their final task of the mission was the installation of a spare arm on Dextre.[8]

1 June STS-134/Endeavour landing

7 June Soyuz-TMA 02M/ ISS Expeditions 28 and 29 launch

8 July STS-135/Atlantis launch

7. Ibid.

8. Ibid.

12 JULY

2011 EVA 9	**Duration:** 6:31
World EVA 349	**Spacecraft/mission:** STS-135
U.S. EVA 242	**Crew:** Christopher Ferguson, Douglas Hurley, Sandra Magnus, Rex Walheim
Shuttle EVA 160	**Spacewalkers:** Michael Fossum, Ronald Garan (ISS Expedition 28 crew members)
	Purpose: Stow failed cooling pump in Atlantis's payload bay; install Robotic Refueling Mission experiment on Dextre platform; install optical mirror to MISSE 8 experiment

The last U.S. spacewalk during the Space Shuttle era was performed by Station residents Fossum and Garan. They moved the Station's failed 1,400-pound (635.04-kilogram) pump, installed two experiments, and repaired a new base for the Canadarm2. The pair used the Canadarm2 to drive the cooling pump to Atlantis's payload bay. There, they bolted and secured the pump for its return to Earth. Next, the spacewalkers installed the Robotic Refueling Mission experiment, which helped investigate methods to refuel space satellites robotically. Garan then deployed the optical mirror, a component of the MISSE 8 that was delayed. With time remaining, the spacewalkers completed get-ahead tasks. They mounted the insulation cover over an area of the PMA-3 that received significant sunshine, rewired obstructing cables under Zarya, and retrieved a cutting device for the following excursion.[9]

21 July STS-135/Atlantis landing

3 AUGUST

2011 EVA 10	**Duration:** 6:23
World EVA 350	**Spacecraft/mission:** International Space Station Expedition 28
Russian EVA 136	**Crew:** Andrey Borisenko (Russian Space Agency), Michael Fossum (NASA), Satoshi Furukawa (JAXA), Ronald Garan (NASA), Alexander Samokutyaev (Russian Space Agency), Sergei Volkov (Russian Space Agency)
Space Station EVA 160	**Spacewalkers:** Sergei Volkov, Alexander Samokutyaev
ISS EVA 55	**Purpose:** Relocate Strela crane from Pirs Docking Compartment to MRM-2 and jettison base thermal cover; install Biorisk-MSN materials exposure experiment; deploy manually 57-pound (25.85-kilogram) free-flying amateur radio satellite known by two names: ARISSat-1 and Radioskaf-V; install BTLS-N monoblock experiment

The crew stowed the ARISSat-1/Radioskaf-V amateur radio satellite in the airlock prior to the EVA for manual deployment during the EVA. After egress, the cosmonauts prepared the satellite for manual

9. "STATUS REPORT: STS-135-09," Amiko Kauderer, editor, accessed 20 June 2014, (*http://www.nasa.gov/mission_pages/shuttle/shuttlemissions/sts135/news/STS-135-09.html*).

deployment. They noticed that one of the two antennas was missing. They decided to delay this task until they received directions from Mission Control. They were given the go-ahead to deploy the satellite in a retrograde direction with one antenna missing. The next task for Volkov and Samokutyaev was the installation of the BTLS-N (Onboard Laser Communications Terminal) monoblock experiment on the universal work platform URM-D. The Strela crane relocation task was cancelled due to the lack of time remaining in the EVA.[10]

15 September Soyuz-TMA 21/ISS Expeditions 27 and 28 landing

13 November Soyuz-TMA 22/ISS Expeditions 29 and 30 launch

21 November Soyuz-TMA 02M/ISS Expeditions 28 and 29 landing

21 December Soyuz-TMA 03M/ISS Expeditions 30 and 31 launch

10. "Cosmonauts Wrap Up Spacewalk," Amiko Kauderer, editor, accessed June 20. 2014, (*http://www.nasa.gov/mission_pages/station/expeditions/expedition28/russian_eva29.html*).

ACRONYMS AND ABBREVIATIONS

ACS	Advanced Camera for Surveys
AERCam	Autonomous Extravehicular Activity Robotic Camera
AMS	Alpha Magnetic Spectrometer
ATV	Automated Transfer Vehicle
BGA1A	Beta Gimbal Assembly 1A
BMRRM	Bearing Motor Roll Ring Module
CAIB	Columbia Accident Investigation Board
CAPCON	Capsule Communicator
CBM	Common Berthing Mechanism
CETA	Crew Equipment Translation Aid
CLPA	Camera Light Pan Tilt Assembly
CNES	Centre National d'Études Spatiale
CNSA	China National Space Administration
COSTAR	Corrective Optics Space Telescope Axial Replacement
CSA	Canadian Space Agency
EAS	Early Ammonia Servicer
ECOMM	Early Communication
EMU	Extravehicular Mobility Unit
ESA	European Space Agency
ESM	Electronics Support Module
ETVCG	External Television Camera Group
EuTEF	European Technology Exposure Facility
ExPRESS	Expedite the Processing of Experiments to the Space Station
FGB	Functional Cargo Block
FPMU	Floating Potential Measurement Unit
FPP	Floating Potential Probe
FRGF	Flight Releasable Grapple Fixture
GATOR	Grappling Adaptor To On-Orbit Railing
GPS	Global Positioning System
HST	Hubble Space Telescope
HUT	Hard Upper Torso
ICC-VLD	Integrated Cargo Carrier-Vertical Light Deployable
ISS	International Space Station
IV	Intravehicular
IVA	Intravehicular Activity
JAXA	Japan Aerospace Exploration Agency
JEM	Japanese Experiment Module
JTVE	JEM Television Electronics
LMC	Lightweight Multi-Purpose Equipment Support Structure Carrier
MAXI	Monitor of All-sky X-ray Image
MBS	Mobile Base System
MISSE	Materials International Space Station Experiment

MPLM	Multi-Purpose Logistics Module
MPAC/SEED	Micro-Particle Capturer and Space Environment Exposure Devices
MRM-1	Mini-Research Module-1
NASDA	National Space Development Agency of Japan
NICMOS	Near Infrared Camera and Multi-Object Spectrometer
NOAX	Non-Oxide Adhesive eXperimental
NSAU	National Space Agency of Ukraine
NTA	Nitrogen Tank Assembly
OBSS	Orbiter Boom Sensor System
OPM	Optical Properties Monitor
ORU	Orbital Replacement Unit
P1	Port One
P3	Port Three
P4	Port Four
P5	Port Five
P6	Port Six
PAS	Payload Attachment System
PCE	Proximity Communications Equipment
PCU	Power Control Unit
PE	Principle Expedition
PMA-1	Pressurized Mating Adapter 1
PMA-2	Pressurized Mating Adapter 2
PMA-3	Pressurized Mating Adapter 3
POA	Payload Orbital Replacement Unit Accommodation
PPP	Portable Power Plant
RMS	Remote Manipulator System
RPCM	Remote Power Control Module
RSA	Russian Federal Space Agency
RSU	Rate Sensor Unit
S0	Starboard Zero
S1	Starboard One
S3	Starboard Three
S4	Starboard Four
S5	Starboard Five
SAFER	Simplified Aid for EVA Rescue
SARJ	Solar Alpha Rotary Joint
SKK	Replaceable Cassette Container
SMDP	Service Module Debris Protection
SOHO	Solar and Heliospheric Observatory
SPD	Spool Positioning Devices
SPDM	Special Purpose Dexterous Manipulator
SPSR	Space Portable Spectroreflectormeter
SSPTS	Space Shuttle Power Transfer System
SSRMS	Space Station Remote Manipulator System
STIS	Space Telescope Imagining Spectrograph
UCCAS	Unpressurized Cargo Carrier Attachment System
UHF	Ultra-High Frequency
USMP-4	United States Microgravity Payload-4
VDU	Video Distribution Unit
VDU	Vynosnaya Dvigatel'naya Ustanovka
WETA	Wireless External Transceiver Assembly
Z1	Zenith One

NASA HISTORY SERIES

REFERENCE WORKS, NASA SP-4000:

Grimwood, James M. *Project Mercury: A Chronology.* NASA SP-4001, 1963.

Grimwood, James M., and Barton C. Hacker, with Peter J. Vorzimmer. *Project Gemini Technology and Operations: A Chronology.* NASA SP-4002, 1969.

Link, Mae Mills. *Space Medicine in Project Mercury.* NASA SP-4003, 1965.

Astronautics and Aeronautics, 1963: Chronology of Science, Technology, and Policy. NASA SP-4004, 1964.

Astronautics and Aeronautics, 1964: Chronology of Science, Technology, and Policy. NASA SP-4005, 1965.

Astronautics and Aeronautics, 1965: Chronology of Science, Technology, and Policy. NASA SP-4006, 1966.

Astronautics and Aeronautics, 1966: Chronology of Science, Technology, and Policy. NASA SP-4007, 1967.

Astronautics and Aeronautics, 1967: Chronology of Science, Technology, and Policy. NASA SP-4008, 1968.

Ertel, Ivan D., and Mary Louise Morse. *The Apollo Spacecraft: A Chronology, Volume I, Through November 7, 1962.* NASA SP-4009, 1969.

Morse, Mary Louise, and Jean Kernahan Bays. *The Apollo Spacecraft: A Chronology, Volume II, November 8, 1962–September 30, 1964.* NASA SP-4009, 1973.

Brooks, Courtney G., and Ivan D. Ertel. *The Apollo Spacecraft: A Chronology, Volume III, October 1, 1964–January 20, 1966.* NASA SP-4009, 1973.

Ertel, Ivan D., and Roland W. Newkirk, with Courtney G. Brooks. *The Apollo Spacecraft: A Chronology, Volume IV, January 21, 1966–July 13, 1974.* NASA SP-4009, 1978.

Astronautics and Aeronautics, 1968: Chronology of Science, Technology, and Policy. NASA SP-4010, 1969.

Newkirk, Roland W., and Ivan D. Ertel, with Courtney G. Brooks. *Skylab: A Chronology.* NASA SP-4011, 1977.

Van Nimmen, Jane, and Leonard C. Bruno, with Robert L. Rosholt. *NASA Historical Data Book, Volume I: NASA Resources, 1958–1968.* NASA SP-4012, 1976; rep. ed. 1988.

Ezell, Linda Neuman. *NASA Historical Data Book, Volume II: Programs and Projects, 1958–1968.* NASA SP-4012, 1988.

Ezell, Linda Neuman. *NASA Historical Data Book, Volume III: Programs and Projects, 1969–1978.* NASA SP-4012, 1988.

Gawdiak, Ihor, with Helen Fedor. *NASA Historical Data Book, Volume IV: NASA Resources, 1969–1978.* NASA SP-4012, 1994.

Rumerman, Judy A. *NASA Historical Data Book, Volume V: NASA Launch Systems, Space Transportation, Human Spaceflight, and Space Science, 1979–1988.* NASA SP-4012, 1999.

Rumerman, Judy A. *NASA Historical Data Book, Volume VI: NASA Space Applications, Aeronautics and Space Research and Technology, Tracking and*

Data Acquisition/Support Operations, Commercial Programs, and Resources, 1979–1988. NASA SP-4012, 1999.

Rumerman, Judy A. *NASA Historical Data Book, Volume VII: NASA Launch Systems, Space Transportation, Human Spaceflight, and Space Science, 1989–1998*. NASA SP-2009-4012, 2009.

Rumerman, Judy A. *NASA Historical Data Book, Volume VIII: NASA Earth Science and Space Applications, Aeronautics, Technology, and Exploration, Tracking and Data Acquisition/Space Operations, Facilities and Resources, 1989–1998*. NASA SP-2012-4012, 2012.

No SP-4013.

Astronautics and Aeronautics, 1969: Chronology of Science, Technology, and Policy. NASA SP-4014, 1970.

Astronautics and Aeronautics, 1970: Chronology of Science, Technology, and Policy. NASA SP-4015, 1972.

Astronautics and Aeronautics, 1971: Chronology of Science, Technology, and Policy. NASA SP-4016, 1972.

Astronautics and Aeronautics, 1972: Chronology of Science, Technology, and Policy. NASA SP-4017, 1974.

Astronautics and Aeronautics, 1973: Chronology of Science, Technology, and Policy. NASA SP-4018, 1975.

Astronautics and Aeronautics, 1974: Chronology of Science, Technology, and Policy. NASA SP-4019, 1977.

Astronautics and Aeronautics, 1975: Chronology of Science, Technology, and Policy. NASA SP-4020, 1979.

Astronautics and Aeronautics, 1976: Chronology of Science, Technology, and Policy. NASA SP-4021, 1984.

Astronautics and Aeronautics, 1977: Chronology of Science, Technology, and Policy. NASA SP-4022, 1986.

Astronautics and Aeronautics, 1978: Chronology of Science, Technology, and Policy. NASA SP-4023, 1986.

Astronautics and Aeronautics, 1979–1984: Chronology of Science, Technology, and Policy. NASA SP-4024, 1988.

Astronautics and Aeronautics, 1985: Chronology of Science, Technology, and Policy. NASA SP-4025, 1990.

Noordung, Hermann. *The Problem of Space Travel: The Rocket Motor*. Edited by Ernst Stuhlinger and J. D. Hunley, with Jennifer Garland. NASA SP-4026, 1995.

Gawdiak, Ihor Y., Ramon J. Miro, and Sam Stueland. *Astronautics and Aeronautics, 1986–1990: A Chronology*. NASA SP-4027, 1997.

Gawdiak, Ihor Y., and Charles Shetland. *Astronautics and Aeronautics, 1991–1995: A Chronology*. NASA SP-2000-4028, 2000.

Orloff, Richard W. *Apollo by the Numbers: A Statistical Reference*. NASA SP-2000-4029, 2000.

Lewis, Marieke, and Ryan Swanson. *Astronautics and Aeronautics: A Chronology, 1996–2000*. NASA SP-2009-4030, 2009.

Ivey, William Noel, and Marieke Lewis. *Astronautics and Aeronautics: A Chronology, 2001–2005*. NASA SP-2010-4031, 2010.

Buchalter, Alice R., and William Noel Ivey. *Astronautics and Aeronautics: A Chronology, 2006*. NASA SP-2011-4032, 2010.

Lewis, Marieke. *Astronautics and Aeronautics: A Chronology, 2007*. NASA SP-2011-4033, 2011.

Lewis, Marieke. *Astronautics and Aeronautics: A Chronology, 2008*. NASA SP-2012-4034, 2012.

Lewis, Marieke. *Astronautics and Aeronautics: A Chronology, 2009*. NASA SP-2012-4035, 2012.

Flattery, Meaghan. *Astronautics and Aeronautics: A Chronology, 2010*. NASA SP-2013-4037, 2014.

MANAGEMENT HISTORIES, NASA SP-4100:

Rosholt, Robert L. *An Administrative History of NASA, 1958–1963*. NASA SP-4101, 1966.

Levine, Arnold S. *Managing NASA in the Apollo Era*. NASA SP-4102, 1982.

Roland, Alex. *Model Research: The National Advisory Committee for Aeronautics, 1915–1958*. NASA SP-4103, 1985.

Fries, Sylvia D. *NASA Engineers and the Age of Apollo*. NASA SP-4104, 1992.

Glennan, T. Keith. *The Birth of NASA: The Diary of T. Keith Glennan*. Edited by J. D. Hunley. NASA SP-4105, 1993.

Seamans, Robert C. *Aiming at Targets: The Autobiography of Robert C. Seamans*. NASA SP-4106, 1996.

Garber, Stephen J., ed. *Looking Backward, Looking Forward: Forty Years of Human Spaceflight Symposium*. NASA SP-2002-4107, 2002.

Mallick, Donald L., with Peter W. Merlin. *The Smell of Kerosene: A Test Pilot's Odyssey*. NASA SP-4108, 2003.

Iliff, Kenneth W., and Curtis L. Peebles. *From Runway to Orbit: Reflections of a NASA Engineer*. NASA SP-2004-4109, 2004.

Chertok, Boris. *Rockets and People, Volume I*. NASA SP-2005-4110, 2005.

Chertok, Boris. *Rockets and People: Creating a Rocket Industry, Volume II*. NASA SP-2006-4110, 2006.

Chertok, Boris. *Rockets and People: Hot Days of the Cold War, Volume III*. NASA SP-2009-4110, 2009.

Chertok, Boris. *Rockets and People: The Moon Race, Volume IV*. NASA SP-2011-4110, 2011.

Laufer, Alexander, Todd Post, and Edward Hoffman. *Shared Voyage: Learning and Unlearning from Remarkable Projects*. NASA SP-2005-4111, 2005.

Dawson, Virginia P., and Mark D. Bowles. *Realizing the Dream of Flight: Biographical Essays in Honor of the Centennial of Flight, 1903–2003*. NASA SP-2005-4112, 2005.

Mudgway, Douglas J. *William H. Pickering: America's Deep Space Pioneer*. NASA SP-2008-4113, 2008.

Wright, Rebecca, Sandra Johnson, and Steven J. Dick. *NASA at 50: Interviews with NASA's Senior Leadership*. NASA SP-2012-4114, 2012.

PROJECT HISTORIES, NASA SP-4200:

Swenson, Loyd S., Jr., James M. Grimwood, and Charles C. Alexander. *This New Ocean: A History of Project Mercury*. NASA SP-4201, 1966; rep. ed. 1999.

Green, Constance McLaughlin, and Milton Lomask. *Vanguard: A History*. NASA SP-4202, 1970; rep. ed. Smithsonian Institution Press, 1971.

Hacker, Barton C., and James M. Grimwood. *On the Shoulders of Titans: A History of Project Gemini*. NASA SP-4203, 1977; rep. ed. 2002.

Benson, Charles D., and William Barnaby Faherty. *Moonport: A History of Apollo Launch Facilities and Operations*. NASA SP-4204, 1978.

Brooks, Courtney G., James M. Grimwood, and Loyd S. Swenson, Jr. *Chariots for Apollo: A History of Manned Lunar Spacecraft*. NASA SP-4205, 1979.

Bilstein, Roger E. *Stages to Saturn: A Technological History of the Apollo/Saturn Launch Vehicles*. NASA SP-4206, 1980 and 1996.

No SP-4207.

Compton, W. David, and Charles D. Benson. *Living and Working in Space: A History of Skylab*. NASA SP-4208, 1983.

Ezell, Edward Clinton, and Linda Neuman Ezell. *The Partnership: A History of the Apollo-Soyuz Test Project*. NASA SP-4209, 1978.

Hall, R. Cargill. *Lunar Impact: A History of Project Ranger*. NASA SP-4210, 1977.

Newell, Homer E. *Beyond the Atmosphere: Early Years of Space Science*. NASA SP-4211, 1980.

Ezell, Edward Clinton, and Linda Neuman Ezell. *On Mars: Exploration of the Red Planet, 1958–1978*. NASA SP-4212, 1984.

Pitts, John A. *The Human Factor: Biomedicine in the Manned Space Program to 1980.* NASA SP-4213, 1985.

Compton, W. David. *Where No Man Has Gone Before: A History of Apollo Lunar Exploration Missions.* NASA SP-4214, 1989.

Naugle, John E. *First Among Equals: The Selection of NASA Space Science Experiments.* NASA SP-4215, 1991.

Wallace, Lane E. *Airborne Trailblazer: Two Decades with NASA Langley's 737 Flying Laboratory.* NASA SP-4216, 1994.

Butrica, Andrew J., ed. *Beyond the Ionosphere: Fifty Years of Satellite Communications.* NASA SP-4217, 1997.

Butrica, Andrew J. *To See the Unseen: A History of Planetary Radar Astronomy.* NASA SP-4218, 1996.

Mack, Pamela E., ed. *From Engineering Science to Big Science: The NACA and NASA Collier Trophy Research Project Winners.* NASA SP-4219, 1998.

Reed, R. Dale. *Wingless Flight: The Lifting Body Story.* NASA SP-4220, 1998.

Heppenheimer, T. A. *The Space Shuttle Decision: NASA's Search for a Reusable Space Vehicle.* NASA SP-4221, 1999.

Hunley, J. D., ed. *Toward Mach 2: The Douglas D-558 Program.* NASA SP-4222, 1999.

Swanson, Glen E., ed. *"Before This Decade Is Out…" Personal Reflections on the Apollo Program.* NASA SP-4223, 1999.

Tomayko, James E. *Computers Take Flight: A History of NASA's Pioneering Digital Fly-By-Wire Project.* NASA SP-4224, 2000.

Morgan, Clay. *Shuttle-Mir: The United States and Russia Share History's Highest Stage.* NASA SP-2001-4225, 2001.

Leary, William M. *"We Freeze to Please": A History of NASA's Icing Research Tunnel and the Quest for Safety.* NASA SP-2002-4226, 2002.

Mudgway, Douglas J. *Uplink-Downlink: A History of the Deep Space Network, 1957–1997.* NASA SP-2001-4227, 2001.

No SP-4228 or SP-4229.

Dawson, Virginia P., and Mark D. Bowles. *Taming Liquid Hydrogen: The Centaur Upper Stage Rocket, 1958–2002.* NASA SP-2004-4230, 2004.

Meltzer, Michael. *Mission to Jupiter: A History of the Galileo Project.* NASA SP-2007-4231, 2007.

Heppenheimer, T. A. *Facing the Heat Barrier: A History of Hypersonics.* NASA SP-2007-4232, 2007.

Tsiao, Sunny. *"Read You Loud and Clear!" The Story of NASA's Spaceflight Tracking and Data Network.* NASA SP-2007-4233, 2007.

Meltzer, Michael. *When Biospheres Collide: A History of NASA's Planetary Protection Programs.* NASA SP-2011-4234, 2011.

CENTER HISTORIES, NASA SP-4300:

Rosenthal, Alfred. *Venture into Space: Early Years of Goddard Space Flight Center.* NASA SP-4301, 1985.

Hartman, Edwin P. *Adventures in Research: A History of Ames Research Center, 1940–1965.* NASA SP-4302, 1970.

Hallion, Richard P. *On the Frontier: Flight Research at Dryden, 1946–1981.* NASA SP-4303, 1984.

Muenger, Elizabeth A. *Searching the Horizon: A History of Ames Research Center, 1940–1976.* NASA SP-4304, 1985.

Hansen, James R. *Engineer in Charge: A History of the Langley Aeronautical Laboratory, 1917–1958.* NASA SP-4305, 1987.

Dawson, Virginia P. *Engines and Innovation: Lewis Laboratory and American Propulsion Technology.* NASA SP-4306, 1991.

Dethloff, Henry C. *"Suddenly Tomorrow Came…": A History of the Johnson Space Center, 1957–1990.* NASA SP-4307, 1993.

Hansen, James R. *Spaceflight Revolution: NASA Langley Research Center from Sputnik to Apollo.* NASA SP-4308, 1995.

Wallace, Lane E. *Flights of Discovery: An Illustrated History of the Dryden Flight Research Center.* NASA SP-4309, 1996.

Herring, Mack R. *Way Station to Space: A History of the John C. Stennis Space Center*. NASA SP-4310, 1997.

Wallace, Harold D., Jr. *Wallops Station and the Creation of an American Space Program*. NASA SP-4311, 1997.

Wallace, Lane E. *Dreams, Hopes, Realities. NASA's Goddard Space Flight Center: The First Forty Years*. NASA SP-4312, 1999.

Dunar, Andrew J., and Stephen P. Waring. *Power to Explore: A History of Marshall Space Flight Center, 1960–1990*. NASA SP-4313, 1999.

Bugos, Glenn E. *Atmosphere of Freedom: Sixty Years at the NASA Ames Research Center*. NASA SP-2000-4314, 2000.

Bugos, Glenn E. *Atmosphere of Freedom: Seventy Years at the NASA Ames Research Center*. NASA SP-2010-4314, 2010. Revised version of NASA SP-2000-4314.

Bugos, Glenn E. *Atmosphere of Freedom: Seventy Five Years at the NASA Ames Research Center*. NASA SP-2014-4314, 2014. Revised version of NASA SP-2000-4314.

No SP-4315.

Schultz, James. *Crafting Flight: Aircraft Pioneers and the Contributions of the Men and Women of NASA Langley Research Center*. NASA SP-2003-4316, 2003.

Bowles, Mark D. *Science in Flux: NASA's Nuclear Program at Plum Brook Station, 1955–2005*. NASA SP-2006-4317, 2006.

Wallace, Lane E. *Flights of Discovery: An Illustrated History of the Dryden Flight Research Center*. NASA SP-2007-4318, 2007. Revised version of NASA SP-4309.

Arrighi, Robert S. *Revolutionary Atmosphere: The Story of the Altitude Wind Tunnel and the Space Power Chambers*. NASA SP-2010-4319, 2010.

GENERAL HISTORIES, NASA SP-4400:

Corliss, William R. *NASA Sounding Rockets, 1958–1968: A Historical Summary*. NASA SP-4401, 1971.

Wells, Helen T., Susan H. Whiteley, and Carrie Karegeannes. *Origins of NASA Names*. NASA SP-4402, 1976.

Anderson, Frank W., Jr. *Orders of Magnitude: A History of NACA and NASA, 1915–1980*. NASA SP-4403, 1981.

Sloop, John L. *Liquid Hydrogen as a Propulsion Fuel, 1945–1959*. NASA SP-4404, 1978.

Roland, Alex. *A Spacefaring People: Perspectives on Early Spaceflight*. NASA SP-4405, 1985.

Bilstein, Roger E. *Orders of Magnitude: A History of the NACA and NASA, 1915–1990*. NASA SP-4406, 1989.

Logsdon, John M., ed., with Linda J. Lear, Jannelle Warren Findley, Ray A. Williamson, and Dwayne A. Day. *Exploring the Unknown: Selected Documents in the History of the U.S. Civil Space Program, Volume I: Organizing for Exploration*. NASA SP-4407, 1995.

Logsdon, John M., ed., with Dwayne A. Day and Roger D. Launius. *Exploring the Unknown: Selected Documents in the History of the U.S. Civil Space Program, Volume II: External Relationships*. NASA SP-4407, 1996.

Logsdon, John M., ed., with Roger D. Launius, David H. Onkst, and Stephen J. Garber. *Exploring the Unknown: Selected Documents in the History of the U.S. Civil Space Program, Volume III: Using Space*. NASA SP-4407, 1998.

Logsdon, John M., ed., with Ray A. Williamson, Roger D. Launius, Russell J. Acker, Stephen J. Garber, and Jonathan L. Friedman. *Exploring the Unknown: Selected Documents in the History of the U.S. Civil Space Program, Volume IV: Accessing Space*. NASA SP-4407, 1999.

Logsdon, John M., ed., with Amy Paige Snyder, Roger D. Launius, Stephen J. Garber, and Regan Anne Newport. *Exploring the Unknown: Selected*

Documents in the History of the U.S. Civil Space Program, Volume V: Exploring the Cosmos. NASA SP-2001-4407, 2001.

Logsdon, John M., ed., with Stephen J. Garber, Roger D. Launius, and Ray A. Williamson. *Exploring the Unknown: Selected Documents in the History of the U.S. Civil Space Program, Volume VI: Space and Earth Science.* NASA SP-2004-4407, 2004.

Logsdon, John M., ed., with Roger D. Launius. *Exploring the Unknown: Selected Documents in the History of the U.S. Civil Space Program, Volume VII: Human Spaceflight: Projects Mercury, Gemini, and Apollo.* NASA SP-2008-4407, 2008.

Siddiqi, Asif A., *Challenge to Apollo: The Soviet Union and the Space Race, 1945–1974.* NASA SP-2000-4408, 2000.

Hansen, James R., ed. *The Wind and Beyond: Journey into the History of Aerodynamics in America, Volume 1: The Ascent of the Airplane.* NASA SP-2003-4409, 2003.

Hansen, James R., ed. *The Wind and Beyond: Journey into the History of Aerodynamics in America, Volume 2: Reinventing the Airplane.* NASA SP-2007-4409, 2007.

Hogan, Thor. *Mars Wars: The Rise and Fall of the Space Exploration Initiative.* NASA SP-2007-4410, 2007.

Vakoch, Douglas A., ed. *Psychology of Space Exploration: Contemporary Research in Historical Perspective.* NASA SP-2011-4411, 2011.

Ferguson, Robert G., *NASA's First A: Aeronautics from 1958 to 2008.* NASA SP-2012-4412, 2013.

Vakoch, Douglas A., ed. *Archaeology, Anthropology, and Interstellar Communication.* NASA SP-2013-4413, 2014.

MONOGRAPHS IN AEROSPACE HISTORY, NASA SP-4500:

Launius, Roger D., and Aaron K. Gillette, comps. *Toward a History of the Space Shuttle: An Annotated Bibliography.* Monographs in Aerospace History, No. 1, 1992.

Launius, Roger D., and J. D. Hunley, comps. *An Annotated Bibliography of the Apollo Program.* Monographs in Aerospace History, No. 2, 1994.

Launius, Roger D. *Apollo: A Retrospective Analysis.* Monographs in Aerospace History, No. 3, 1994.

Hansen, James R. *Enchanted Rendezvous: John C. Houbolt and the Genesis of the Lunar-Orbit Rendezvous Concept.* Monographs in Aerospace History, No. 4, 1995.

Gorn, Michael H. *Hugh L. Dryden's Career in Aviation and Space.* Monographs in Aerospace History, No. 5, 1996.

Powers, Sheryll Goecke. *Women in Flight Research at NASA Dryden Flight Research Center from 1946 to 1995.* Monographs in Aerospace History, No. 6, 1997.

Portree, David S. F., and Robert C. Trevino. *Walking to Olympus: An EVA Chronology.* Monographs in Aerospace History, No. 7, 1997.

Logsdon, John M., moderator. *Legislative Origins of the National Aeronautics and Space Act of 1958: Proceedings of an Oral History Workshop.* Monographs in Aerospace History, No. 8, 1998.

Rumerman, Judy A., comp. *U.S. Human Spaceflight: A Record of Achievement, 1961–1998.* Monographs in Aerospace History, No. 9, 1998.

Portree, David S. F. *NASA's Origins and the Dawn of the Space Age.* Monographs in Aerospace History, No. 10, 1998.

Logsdon, John M. *Together in Orbit: The Origins of International Cooperation in the Space Station.* Monographs in Aerospace History, No. 11, 1998.

Phillips, W. Hewitt. *Journey in Aeronautical Research: A Career at NASA Langley Research Center.* Monographs in Aerospace History, No. 12, 1998.

Braslow, Albert L. *A History of Suction-Type Laminar-Flow Control with Emphasis on Flight Research.* Monographs in Aerospace History, No. 13, 1999.

Logsdon, John M., moderator. *Managing the Moon Program: Lessons Learned from Apollo.* Monographs in Aerospace History, No. 14, 1999.

Perminov, V. G. *The Difficult Road to Mars: A Brief History of Mars Exploration in the Soviet Union.* Monographs in Aerospace History, No. 15, 1999.

Tucker, Tom. *Touchdown: The Development of Propulsion Controlled Aircraft at NASA Dryden.* Monographs in Aerospace History, No. 16, 1999.

Maisel, Martin, Demo J. Giulanetti, and Daniel C. Dugan. *The History of the XV-15 Tilt Rotor Research Aircraft: From Concept to Flight.* Monographs in Aerospace History, No. 17, 2000. NASA SP-2000-4517.

Jenkins, Dennis R. *Hypersonics Before the Shuttle: A Concise History of the X-15 Research Airplane.* Monographs in Aerospace History, No. 18, 2000. NASA SP-2000-4518.

Chambers, Joseph R. *Partners in Freedom: Contributions of the Langley Research Center to U.S. Military Aircraft of the 1990s.* Monographs in Aerospace History, No. 19, 2000. NASA SP-2000-4519.

Waltman, Gene L. *Black Magic and Gremlins: Analog Flight Simulations at NASA's Flight Research Center.* Monographs in Aerospace History, No. 20, 2000. NASA SP-2000-4520.

Portree, David S. F. *Humans to Mars: Fifty Years of Mission Planning, 1950–2000.* Monographs in Aerospace History, No. 21, 2001. NASA SP-2001-4521.

Thompson, Milton O., with J. D. Hunley. *Flight Research: Problems Encountered and What They Should Teach Us.* Monographs in Aerospace History, No. 22, 2001. NASA SP-2001-4522.

Tucker, Tom. *The Eclipse Project.* Monographs in Aerospace History, No. 23, 2001. NASA SP-2001-4523.

Siddiqi, Asif A. *Deep Space Chronicle: A Chronology of Deep Space and Planetary Probes, 1958–2000.* Monographs in Aerospace History, No. 24, 2002. NASA SP-2002-4524.

Merlin, Peter W. *Mach 3+: NASA/USAF YF-12 Flight Research, 1969–1979.* Monographs in Aerospace History, No. 25, 2001. NASA SP-2001-4525.

Anderson, Seth B. *Memoirs of an Aeronautical Engineer: Flight Tests at Ames Research Center: 1940–1970.* Monographs in Aerospace History, No. 26, 2002. NASA SP-2002-4526.

Renstrom, Arthur G. *Wilbur and Orville Wright: A Bibliography Commemorating the One-Hundredth Anniversary of the First Powered Flight on December 17, 1903.* Monographs in Aerospace History, No. 27, 2002. NASA SP-2002-4527.

No monograph 28.

Chambers, Joseph R. *Concept to Reality: Contributions of the NASA Langley Research Center to U.S. Civil Aircraft of the 1990s.* Monographs in Aerospace History, No. 29, 2003. NASA SP-2003-4529.

Peebles, Curtis, ed. *The Spoken Word: Recollections of Dryden History, The Early Years.* Monographs in Aerospace History, No. 30, 2003. NASA SP-2003-4530.

Jenkins, Dennis R., Tony Landis, and Jay Miller. *American X-Vehicles: An Inventory—X-1 to X-50.* Monographs in Aerospace History, No. 31, 2003. NASA SP-2003-4531.

Renstrom, Arthur G. *Wilbur and Orville Wright: A Chronology Commemorating the One-Hundredth Anniversary of the First Powered Flight on December 17, 1903.* Monographs in Aerospace History, No. 32, 2003. NASA SP-2003-4532.

Bowles, Mark D., and Robert S. Arrighi. *NASA's Nuclear Frontier: The Plum Brook Research Reactor.* Monographs in Aerospace History, No. 33, 2004. NASA SP-2004-4533.

Wallace, Lane, and Christian Gelzer. *Nose Up: High Angle-of-Attack and Thrust Vectoring Research at NASA Dryden, 1979–2001.* Monographs in Aerospace History, No. 34, 2009. NASA SP-2009-4534.

Matranga, Gene J., C. Wayne Ottinger, Calvin R. Jarvis, and D. Christian Gelzer. *Unconventional, Contrary, and Ugly: The Lunar Landing Research Vehicle.* Monographs in Aerospace History, No. 35, 2006. NASA SP-2004-4535.

McCurdy, Howard E. *Low-Cost Innovation in Spaceflight: The History of the Near Earth Asteroid Rendezvous (NEAR) Mission*. Monographs in Aerospace History, No. 36, 2005. NASA SP-2005-4536.

Seamans, Robert C., Jr. *Project Apollo: The Tough Decisions*. Monographs in Aerospace History, No. 37, 2005. NASA SP-2005-4537.

Lambright, W. Henry. *NASA and the Environment: The Case of Ozone Depletion*. Monographs in Aerospace History, No. 38, 2005. NASA SP-2005-4538.

Chambers, Joseph R. *Innovation in Flight: Research of the NASA Langley Research Center on Revolutionary Advanced Concepts for Aeronautics*. Monographs in Aerospace History, No. 39, 2005. NASA SP-2005-4539.

Phillips, W. Hewitt. *Journey into Space Research: Continuation of a Career at NASA Langley Research Center*. Monographs in Aerospace History, No. 40, 2005. NASA SP-2005-4540.

Rumerman, Judy A., Chris Gamble, and Gabriel Okolski, comps. *U.S. Human Spaceflight: A Record of Achievement, 1961–2006*. Monographs in Aerospace History, No. 41, 2007. NASA SP-2007-4541.

Peebles, Curtis. *The Spoken Word: Recollections of Dryden History Beyond the Sky*. Monographs in Aerospace History, No. 42, 2011. NASA SP-2011-4542.

Dick, Steven J., Stephen J. Garber, and Jane H. Odom. *Research in NASA History*. Monographs in Aerospace History, No. 43, 2009. NASA SP-2009-4543.

Merlin, Peter W. *Ikhana: Unmanned Aircraft System Western States Fire Missions*. Monographs in Aerospace History, No. 44, 2009. NASA SP-2009-4544.

Fisher, Steven C., and Shamim A. Rahman. *Remembering the Giants: Apollo Rocket Propulsion Development*. Monographs in Aerospace History, No. 45, 2009. NASA SP-2009-4545.

Gelzer, Christian. *Fairing Well: From Shoebox to Bat Truck and Beyond, Aerodynamic Truck Research at NASA's Dryden Flight Research Center*. Monographs in Aerospace History, No. 46, 2011. NASA SP-2011-4546.

Arrighi, Robert. *Pursuit of Power: NASA's Propulsion Systems Laboratory No. 1 and 2*. Monographs in Aerospace History, No. 48, 2012. NASA SP-2012-4548.

Goodrich, Malinda K., Alice R. Buchalter, and Patrick M. Miller, comps. *Toward a History of the Space Shuttle: An Annotated Bibliography, Part 2 (1992–2011)*. Monographs in Aerospace History, No. 49, 2012. NASA SP-2012-4549.

Gelzer, Christian. *The Spoken Word III: Recollections of Dryden History; The Shuttle Years*. Monographs in Aerospace History, No. 52, 2013. NASA SP-2013-4552.

Ross, James C. *NASA Photo One*. Monographs in Aerospace History, No. 53, 2013. NASA SP-2013-4553.

Launius, Roger D. *Historical Analogs for the Stimulation of Space Commerce*. Monographs in Aerospace History, No 54, 2014. NASA SP-2014-4554.

Buchalter, Alice R., and Patrick M. Miller, comps. *The National Advisory Committee for Aeronautics: An Annotated Bibliography*. Monographs in Aerospace History, No. 55, 2014. NASA SP-2014-4555.

Chambers, Joseph R., and Mark A. Chambers. *Emblems of Exploration: Logos of the NACA and NASA*. Monographs in Aerospace History, No. 56, 2015. NASA SP-2015-4556.

ELECTRONIC MEDIA, NASA SP-4600:

Remembering Apollo 11: The 30th Anniversary Data Archive CD-ROM. NASA SP-4601, 1999.

Remembering Apollo 11: The 35th Anniversary Data Archive CD-ROM. NASA SP-2004-4601, 2004. This is an update of the 1999 edition.

The Mission Transcript Collection: U.S. Human Spaceflight Missions from Mercury Redstone 3 to Apollo 17. NASA SP-2000-4602, 2001.

Shuttle-Mir: The United States and Russia Share History's Highest Stage. NASA SP-2001-4603, 2002.

U.S. Centennial of Flight Commission Presents Born of Dreams—Inspired by Freedom. NASA SP-2004-4604, 2004.

Of Ashes and Atoms: A Documentary on the NASA Plum Brook Reactor Facility. NASA SP-2005-4605, 2005.

Taming Liquid Hydrogen: The Centaur Upper Stage Rocket Interactive CD-ROM. NASA SP-2004-4606, 2004.

Fueling Space Exploration: The History of NASA's Rocket Engine Test Facility DVD. NASA SP-2005-4607, 2005.

Altitude Wind Tunnel at NASA Glenn Research Center: An Interactive History CD-ROM. NASA SP-2008-4608, 2008.

A Tunnel Through Time: The History of NASA's Altitude Wind Tunnel. NASA SP-2010-4609, 2010.

CONFERENCE PROCEEDINGS, NASA SP-4700:

Dick, Steven J., and Keith Cowing, eds. *Risk and Exploration: Earth, Sea and the Stars.* NASA SP-2005-4701, 2005.

Dick, Steven J., and Roger D. Launius. *Critical Issues in the History of Spaceflight.* NASA SP-2006-4702, 2006.

Dick, Steven J., ed. *Remembering the Space Age: Proceedings of the 50th Anniversary Conference.* NASA SP-2008-4703, 2008.

Dick, Steven J., ed. *NASA's First 50 Years: Historical Perspectives.* NASA SP-2010-4704, 2010.

SOCIETAL IMPACT, NASA SP-4800:

Dick, Steven J., and Roger D. Launius. *Societal Impact of Spaceflight.* NASA SP-2007-4801, 2007.

Dick, Steven J., and Mark L. Lupisella. *Cosmos and Culture: Cultural Evolution in a Cosmic Context.* NASA SP-2009-4802, 2009.

Dick, Steven J. *Historical Studies in the Societal Impact of Spaceflight.* NASA SP-2015-4803, 2015.

INDEX

D

T

U